Allan Pollock

A Botanical Index

To all the medicinal plants, barks, roots, seeds and flowers usually kept by

druggists

Allan Pollock

A Botanical Index

To all the medicinal plants, barks, roots, seeds and flowers usually kept by druggists

ISBN/EAN: 9783742833655

Manufactured in Europe, USA, Canada, Australia, Japa

Cover: Foto ©berggeist007 / pixelio.de

Manufactured and distributed by brebook publishing software
(www.brebook.com)

Allan Pollock

A Botanical Index

A

BOTANICAL INDEX

TO ALL THE

MEDICINAL PLANTS,

BARKS, ROOTS, SEEDS AND FLOWERS USUALLY
KEPT BY DRUGGISTS,

ARRANGED IN ALPHABETICAL ORDER, WITH THEIR OFFICINAL
AND COMMON NAMES.

BY

ALLAN POLLOCK,

NEW EDITION, REVISED AND ENLARGED.

NEW YORK:
ALLAN POLLOCK.
1873.

Francis Hart & Co., Printers, 12 and 14 College Place. N. Y.

ERRATA.

Put Aletris farinosa in italic, p. 44.

Nabulus Albus is incorrectly spelt on pp. 29. 90, 106, 107,

Change the *Officinal* of Zanthoriza to Zanthoriza apiofolia, p. 137.

         ''   of Parsley Yellow Root to   ''

INTRODUCTION.

Feeling the need of a work of this description in the daily labors of his store, the Author was induced to commence the present compilation. At that time he had no idea of its publication, and it was only at the urgent request of many of his friends that he was finally induced to give his consent.

In the preparation of this work, he has endeavored to obtain not only all the Local or Common Names in use in the various sections of the United States, but also in Great Britain, and he trusts that but few, if any, have been overlooked.

In Part I, the Local Name generally adopted and used is put in *Italic*. In Part II, where there is more than one Botanical Name for the same Local Name, the correct one is italicised to distinguish it from the others.

The compilation has been prepared upon strictly Botanical Rules, and as far as the Author can ascertain from those who have critically examined the work, and who are fully competent to judge, it is accurate.

The result of his labors is now submitted to the Pharmacists of the United States, hoping that it may meet with their approbation, and be of service to them.

Respectfully,

ALLAN POLLOCK.

New-York, September, 1872.

INTRODUCTION TO THE SECOND EDITION.

THE *first edition* of this work having been disposed of and the demand constantly increasing, the compiler has given it a *thorough revision;* corrected such errors as were in it, and made considerable additions. He now presents the Second Edition, hoping that it will be found satisfactory in every respect; at the same time he would tender thanks for the very liberal encouragement bestowed on the publication.

ALLAN POLLOCK.

New York, July 1873.

PART I.

Botanical or Officinal Names.	Local or Common Names.
ACHILLEA MILLEFOLIUM......	*Yarrow*, Milfoil, Thousand Leaf.
ACONITUM NAPELLUS............	*Aconite*, Monkshood, Wolfsbane
ACORUS CALAMUS...................	*Flag Root*, Calamus, Sweet Flag Sweet Rush, Myrtle Flag.
ACTÆA ALBA..............	*White Cohosh*, White Beads, Noah's Ark, Necklace Weed, White Baneberry.
" RUBRA.............	*Red Cohosh*, Baneberry.
ADIANTUM PEDATUM	*Maiden Hair*, Rock Fern.
ÆSCULUS GLABRA.........	*Ohio Buckeye*, Horse Chestnut.
" HIPOCASTANUM...	*English Horse Chestnut.*
AGATHOTES CHIRAYITA....	*Chiretta*, Chirayta, Creata, Cherayita, Bitter Stick.
AGAVE VIRGINICA....................	*Rattlesnake's Master*, False Aloe, Aloe Root.
AGRIMONIA EUPATORIA.......	*Common Agrimony*, Cockle Burr, Stick Weed.
AJUGA CHAMÆPITHYS.............	*Yellow Bugle*, Ground Pine.
" REPTANS....................	*Common Bugle.*
" PYRAMIDALIS..............	" "

BOTANICAL OR OFFICINAL NAMES.	LOCAL OR COMMON NAMES.
ALETRIS FARINOSA...............	*Star Grass*, Star Root, Blazing Star, Mealy Starwort, True Unicorn Root, Bitter Grass, Devil's-bit, Aloe Root, Colic Root, Ague Root, Crowcorn, Black Root, Ague Grass.
ALISMA PLANTAGO..............	*Water Plantain.*
ALNUS RUBRA......................	*Red Alder*, Tag Alder, Swamp Alder, Smooth Alder.
"　　SERRULATA	"　　　　"　　　　"
ALTHÆA OFFICINALIS...........	*Marsh Mallows*, Althea, Bismalva, Hibiscus, Ibiscus, Mortification Root.
"　　ROSEA	Hollyhock.
AMYGDALUS PERSICA......	*Peach Leaves*, Nectarine.
AMARANTHUS HYPO- } CHONDRIACUS............... }	*Prince's Feather*, Red Cockscomb, Pilewort, Amaranth,
"　　MELANCHOLICUS	*Love-lies-bleeding.*
AMBROSIA ARTEMISÆFOLIA..	*Rag Weed*, Roman Wormwood, Hog Weed.
ANACYCLUS PYRETHRUM...	*Pelitory Root*, Pelitory of Spain, Pyrethrum Root.
ANAGALLIS ARVENSIS	*Scarlet Pimpernel*, Poor-man's Weather-glass.

BOTANICAL OR OFFICINAL NAMES.	LOCAL OR COMMON NAMES.
ANCHUSA TINCTORIA.............	*Alkanet*, Bugloss, Dyers' Alkanet.
ANEMONE PRATENSIS......	*Meadow Anemone*, Pasque Flower, Wind Flower.
" PULSATILLA.....	*Pulsatilla.*
ANETHUM GRAVEOLENS...	*Dill Seed*, Garden Dill, Dilly.
ANGELICA ATROPURPUREA..	*Purple Angelica*, High Angelica, Wild Angelica, Wild Archangel, Masterwort.
" ,ARCHANGELICA...	*Wild Archangel.*
" SYLVESTRIS......	" *Angelica.*
ANGUSTURA.............	*Angustura Bark.*
ANTIRRHINUM LINARIA......	*Toad Flax*, Yellow Toad Flax, Linaria, Snap Dragon.
ANTHEMIS COTULA...............	*May Weed*, Dog Fennel, Wild Chamomile, Stinking Chamomile, May Weed Flower, Dill Weed, Dilly, Field Weed, Mathen, Flake.
" NOBILIS...............	*Roman Chamomile*, Low Chamomile.
APIUM PETROSELINUM......	*Parsley Leaves*, Seed, and Root.
APLECTRUM HYEMALE.....	*Adam and Eve*, Putty Root, Chicken's Toes, Fever Root.
APOCYNUM ANDROSÆMI-FOLIUM......................... }	*Bitter Root*, Dogsbane, Bitter Dogsbane, Fever-twig, Flytrap, Wandering Milk Weed, Ipecac Milk, Honey Bloom, Catch-fly.
APOCYNUM CANNABINUM......	*Indian Hemp*, Black Hemp.

Botanical or Official Names.	Local or Common Names.
ARALIA HISPIDA...................	*Dwarf Elder*, Wild Elder, Bristly Aralia, Bristle Stem. Sarsaparilla.
" RACEMOSA.................	*Spikenard*, Petty-morel, Life of Man, Spignet.
" SPINOSA..................	*Angelica Tree Bark*, Toothache Bush, Hercules' Club.
" NUDICAULIS..............	*Wild Sarsaparilla* or American, Prickly Ash, Suterberry, Prickly Elder, Shotbush, Yellow Wood, Sea Ash.
ARCTOSTAPHYLOS UVA URSI	*Uva Ursi*, Bearberry, Wild Cranberry, Bear's Whortleberry, Mountain Box, Redberry, Snake Weed, White Sanicle, Snargel, Mountain Cranberry, Upland Cranberry.
ARISTOLOCHIA SERPEN-TARIA	*Virginia Snake Root*, Serpentaria, Snake Root, White Sanicle, Sangrel, Snargel, White Snake Root, Birthwort.
ARNICA MONTANA................	*Arnica*, Leopard-bane, Mountain Tobacco.
ARTANTHE ELONGATA..........	*Matico*.

BOTANICAL OR OFFICINAL NAMES.	LOCAL OR COMMON NAMES.
ARTEMISIA ABROTANUM······	*Southern Wood,* Old Man, Boy's Love, Lad's Love.
" ABSINTHIUM . ..	*Wormwood.*
" VULGARIS.......	*Mugwort,* Wild Wormwood, Yerba Santa.
ARUM TRIPHYLLUM........	*Wild Turnip,* Indian Turnip, Pepper Turnip, Bog Onion, Marsh Turnip, Dragon's Root, Dragon's Turnip, Jack in the Pulpit, Canada Turnip.
" MACULATUM........	*Wake Robin,* Cookoo's Pint.
ASARUM CANADENSE.......	*Wild Ginger,* Indian Ginger, Canada Ginger, Wild Turnip, Kidney-leaved Asarabacca, Canada Snake Root, Heart Root, Coltsfoot, Coltsfoot Snake Root, False Coltsfoot, Vermont Snake Root, Heart Snake Root.
*ASCLEPIAS GARCIANA or } MATA PERRO.................. }	*Cundurango,* Condor Vine, Mata Perro.
ASCLEPIAS INCARNATA...........	*Rose-colored Silk Weed,* Swamp Milk Weed Root, White Indian Hemp, Flesh-colored Asclepias.
" SYRIACA	*Milk Weed Root,* Silk Weed Root.

* No Officinal Name has yet been given to *Cundurango* by the Botanical writers.

BOTANICAL OR OFFICINAL NAMES.	LOCAL OR COMMON NAMES.
ASCLEPIAS TUBEROSA	*Pleurisy Root*, Butterfly Weed, Flux Root, White Flux Root, Colic Root, Windwort Root, White Root, Canada Root, Orange Swallow Root, Orange Apocynum, Swallow-wort, Windwort, Tuber Root, White Tuberous-rooted Swallow-wort.
ASPIDIUM FILIX MAS..	*Male Fern*, Golden Fern, Male Shield Fern.
ASPLENIUM ANGUSTIFOLIUM	*Swamp Spleenwort.*
" FILIX FŒMINA..	*Female Fern*, Backache-brake.
ASTER PUNICEUS	*Frost Weed*, Red-stalked Aster, Cold Water Root, Cocash Root, Meadow Scabish.
ATROPA BELLADONNA	*Belladonna*, Deadly Nightshade, Dwale.
AURANTII CORTEX	*Orange Peel.*
BAPTISIA TINCTORIA	*Wild Indigo*, Indigo Weed, Indigo Broom, Yellow Broom, Horse-fly Weed, Indigofera, Rattle Bush.
BAROSMA CRENATA	*Buchu Leaves.*

BOTANICAL OR OFFICINAL NAMES.	LOCAL OR COMMON NAMES.
BENZOIN ODORIFERUM...... (See *Laurus Benzoin.*)	*Fever Bush*, Fever Wood, Spice Wood, Spice Bush, Spice Berry, Benjamin Bush, Snap Wood, Wild Allspice.
BERBERIS VULGARIS........	*Barberry*, Pipperidge.
BETULA LENTA	*Birch Bark*, Sweet Birch, Cherry Birch, Mahogany Birch, Spice Birch.
BIGNONIA LEUCOXILON.....	*Trumpet Weed.*
BORAGO OFFICINALIS.......	*Borage*, Bugloss, Burrage.
CALAMUS AROMATICUS...... (See *Acorus Calamus.*)	*Flag Root*, Calamus, Sweet Flag, Sweet Rush.
CALENDULA OFFICINALIS....	*Marigold.*
CANELLA ALBA............	*White Canella*, Wild Cinnamon.
CANNABIS INDICA............	*Foreign Indian Hemp*, Cannabis, Ganga.
CAPSICUM ANNUUM.........	*Cayenne Pepper*, Red Pepper, Bird Pepper, Cockspur Pepper, Guinea Pods, Tochillies.
CARTHAMUS TINCTORIUS.....	*Safflower*, Dyers' Saffron, Bastard Saffron, American Saffron.
CASCARILLA............	*Cascarilla Bark.*
CASSIA ACUTIFOLIA............	*Senna*, Indian Senna, Locust Plant.

BOTANICAL OR OFFICINAL NAMES.	LOCAL OR COMMON NAMES.
CASSIA MARILANDICA	*American Senna,* Wild Senna, Maryland Cassia, Locust Plant.
CAULOPHYLLUM THA- LICTROIDES......... } (See *Leontice Thalictroides.*)	*Blue Cohosh,* Leontice, Pappoose Root, Squaw Root, Blueberry Root.
CEANOTHUS AMERICANA...	*Jersey Tea,* Bohea, Red Root Bark, Wild Snow Ball.
CELASTRUS SCANDENS. ...	*Bittersweet Bark,* Climbing Bittersweet, Bittersweet Staff Tree, False Bittersweet, Shrubby Bittersweet, Fevertwig, Wax Work, Staff Vine, Staff Vine Tree, Climbing Orange Root.
CENTAUREA AMERICANA...	*Great American Century,* Succory, Rose Pink, Bitter Clover.
" BENEDICTA....	*Spotted Cardus,* Blessed Thistle, Spotted Thistle, Cursed Thistle, Holy Thistle, Star Thistle, Lovely Thistle, Thistle Root, Cardus Plant.
CEPHÆLIS IPECACUANHA...	*Ipecac.*
CETRARIA ISLANDICA......	*Iceland Moss.*
CHELIDONIUM MAJUS.....	*Celandine,* Calandine, Tetterwort.
CHELONE GLABRA...	*Balmony,* Snake Head, Bitterherb, Turtle Bloom, Salt Rheum Weed, Fishmouth, Turtle Head, Broomshell Flower.

BOTANICAL OR OFFICINAL NAMES.	LOCAL OR COMMON NAMES.
CHENOPODIUM ANTHEL- MINTICUM	*Wormseed,* Jerusalem Oak, Stinking Weed, Worm Goosefoot, Jesuits' Tea.
CHIMAPHILA MACULATA...	*White Leaf,* Pyrola.
" UMBELLATA..	*Pipsissewa,* Prince's Pine, Ground Leaf, Ground Holly, American Wintergreen, King's Cure, Herb de Paigne, White Leaf, Noble Pine, Rheumatism- weed.
CHRYSANTHEMUM PAR- THENIUM (See *Pyrethrum Parthenium.)*	*Feverfew,* Featherfew, Chrysan- themum.
CICUTA MACULATA	*American Water Hemlock,* Spot- ted Cowbane, Beaver Poison, Musquash Root.
" VIROSA	*Water Hemlock,* Water Parsnip, Cowbane, Snake Weed, Long- leaved Water Parsnip, Poison Hemlock.
CIMICIFUGA RACEMOSA.... (See *Macrotys Racemosa.)*	*Black Snake Root,* Black Co- hosh, Deer Weed, Rattle Root, Rich Weed, Squaw Root, Meadow Bloom, Bug- bane.
CINCHONA OFFICINALIS	*Peruvian Bark,* Red Bark, San- icle, Crown Bark, Jesuit's Bark.
CISSAMPELOS PAREIRA	*Pareira Brava,* Ice Vine, Velvet Leaf.

BOTANICAL OR OFFICINAL NAMES.	LOCAL OR COMMON NAMES.
CISTUS CANADENSIS................	*Rock Rose,* Holly Rose, Frost Plant.
CLEMATIS VIRGINICA.......	*Virgin's Bower,* Traveller's Joy.
COCCULUS PALMATUS .. ⎱ (See *Fraseri Walteri.*) ⎰	*Columbo,* Columba, Columbia, Marietta Columbia, Pyramid Flower, Indian Lettuce.
COCHLEARIA ARMORACIA..	*Horse Radish.*
" OFFICINALIS..	*Scurvy Grass.*
COLCHICUM AUTUMNALE...	*Colchicum,* Meadow Saffron.
COLLINSONIA CANADENSIS..	*Horse Weed,* Horse Balm, Ox Balm, Rich Weed, Heal-all, Knob's Grass, Knob Root, Knot Root, Stone Root, Hardhack, Canadian Snake Root, Canada Snake Root, Woundwort.
COMPTONIA ASPLENIFOLIA..	*Sweet Fern,* Sweet Bush, Fern Gale, Sweet Fern Bush, Shrubby Sweet Fern, Spleenwort Bush, Sweet Ferry, Spleenwort-leaved Gale, Meadow Fern.
CONIUM MACULATUM..... ⎱ (See *Cicuta Maculata.*) ⎰	*Water Hemlock,* Poison Hemlock, Cicuta, Poison Parsley, Beaver Poison, Musquash Root, Spotted Cowbane.
CONVALLARIA MAJALIS....	*Lily of the Valley.*
" MULTIFLORA	*Solomon's Seal.*

BOTANICAL OR OFFICINAL NAMES.	LOCAL OR COMMON NAMES.
CONVOLVULUS PANDU-RATUS...............	*Bind Weed*, Man Root, Wild Scammony, Wild Jalap, Man in the Ground, Man in the Earth, Fiddle-leaved Bind Weed, Hog Potato, Mecha-mech, Mecoama, Wild Rhu-barb, Kusauda, Wild Potato, Man of the Earth.
COPTIS TRIFOLIA....................	*Gold Thread*, Mouth Root, Yel-low Root, Canker Root.
CORALLORHIZA ODON-TORHIZA	*Coral Root*, Crawley, Dragon's Claw, Coral Teeth.
CORALLORHIZA HYEMALE (See *Aplectrum Hyemale.*)	*Adam and Eve*, Putty Root, Chicken's Toes, Fever Root.
CORNUS CIRCINATA...............	*Round-leaved Dogwood* or Cor-nel, Cornea, Green Osier, Swamp Sassafras.
FLORIDA...	*Dogwood*, Boxwood, Large Flowering Cornel, Flowering Dogwood, Dog Tree, New England Boxwood, Great Flowering Dogwood, Florida Dogwood, Male Virginian Dogwood, Box Tree, Bud-wood.

BOTANICAL OR OFFICINAL NAMES.	LOCAL OR COMMON NAMES.
CORNUS SERICEA	*Swamp Dogwood,* Willow Rose, Rose Willow, Red Osier, Red Willow, Red Rod, Female Dogwood, Silky-leaved Dogwood, Blue-berried Dogwood, Blue-berried Cornus, American Red Cornel, Silky Cornel.
CORYDALIS FORMOSA	*Turkey Corn,* Turkey Pea, Fumitory, Stagger Weed, Dielytra.
CROCUS SATIVUS	*Saffron,* Spanish Saffron, Foreign Saffron.
CROTON ELEUTERIA	*Cascarilla Bark,* Sea-side Balsam.
CUCUMIS COLOCYNTHIS	*Colocynth,* Bitter Apple, Bitter Cucumber.
CUCURBITA CITRULLUS	*Watermelon Seed.*
" PEPO	*Pumpkin Seed.*
CUNILA MARIANA	*American Dittany,* Mountain Dittany, Mint-leaved Cunila, Maryland Cunila, Wild Basil, Stone Mint.
CURCUMA LONGA	*Turmeric,* Curcuma.
CYDONIA VULGARIS	*Quince Seed.*

BOTANICAL OR OFFICINAL NAMES.	LOCAL OR COMMON NAMES.
CYNOGLOSSUM OFFICINALIS	*Hound's Tongue.*
CYPRIPEDIUM PUBESCENS..	*Lady's Slipper,* Large Yellow Lady's Slipper, Nerve Root, Water Nerve Root, Nervine, Moccasin Root, Umbil Root, Noah's Ark, Indian Shoe, Bleeding Heart, American Valerian, Pine Tulip.
CYTISUS SCOPARIUS...............	*Broom Herb.*
DAUCUS CAROTA...................	*Garden Carrot,* Wild Carrot, Bee's Nest, Bird's Nest.
DAPHNE MEZEREUM..............	*Mezereum,* Olive Spurge.
DATURA STRAMONIUM...........	*Jamestown Weed,* Jimpson Weed, Devil's Apple, Stink Weed, Stramonia, Apple of Peru, Thorn Apple.
DELPHINIUM CONSOLIDA....	*Field Larkspur,* Branching Larkspur.
" STAPHISAGRIA	*Stavesacre,* Larkspur.
DIGITALIS PURPUREA...........	*Digitalis,* Foxglove, Purple Glove, Fairy's Glove.
DIOSCOREA VILLOSA.............	*Wild Yam,* Colic Root, China Root, Devil's Bones.
DORSTENIA CONTRAYERVA...	*Contrayerva,* Counter Poison, Antidote.
DRACOCEPHALUM CA- NARIENSE........................	*Dragon's Head,* Sweet Balm, Balm of Gilead Herb.

BOTANICAL OR OFFICINAL NAMES.	LOCAL OR COMMON NAMES.
DRACONTIUM FŒTIDUM SYMPLOCARPUS FŒTIDUS ICTODES FŒTIDA............. }	· *Skunk Cabbage*, Skunk Weed, Collard, Cow Collard, Polecat Collard, Meadow Cabbage, Swamp Cabbage, Irish Cabbage, Itch Weed, Skoka, Polecat Weed, Fœtid Hellebore, Stinking Pothos, Bear's-leaf, Bear's-foot, Poke, Biornretter, Beerenwortel, Bonsemkrenia, Byorn-blad.
EPIGÆA REPENS..................	*Trailing Arbutus*, Gravel Plant, Gravel Laurel, Ground Laurel, May Flower, Winter Pink, Gravel Weed. Mountain Pink.
EPILOBIUM ANGUSTIFOLIUM	*Wickup*, Willow Herb.
" SPICATUM..........	" "
EPIPHEGUS AMERICANUS......	*Cancer Root*, Beech Drop, Broom Rape.
ERECHTHITES HIERACI- FOLIA............................ }	*Fire Weed.* (See *Senecio Hieracifolia.*)
ERIGERON CANADENSE.........	*Canada Fleabane*, Horse Weed, Butter Horse Weed, Blood Stanch, Butter Weed, Mare's Tail, Colt's Tail, Scabious, Pride Weed.
" HETEROPHYLLUM	*Various-leaved Fleabane.*

BOTANICAL OR OFFICINAL NAMES.	LOCAL OR COMMON NAMES.
ERIGERON PHILADELPHICUM... ERIGERON PURPUREUM....	*Philadelphia Fleabane,* Cocash, Sweet Scabious, Squaw Weed, Skevish, Narrow-rayed Robins Plantation, Scavish.
ERYNGIUM AQUATICUM........	*Water Eryngo,* Button Snake Root, Rattlesnake's Master, · Corn Snake Root.
ERYTHRÆA CENTAURIUM..	*European Century.*
ERYTHRONIUM AMERICANUM..............	*Erythronium,* Yellow Adder's Tongue, Yellow Snake Leaf, Dog-tooth Violet, Rattlesnake Violet.
EUCALYPTUS GLOBULUS....	*Australian Blue Gum.*
EUONYMUS ATROPURPUREUS........................	*Waahoo,* Burning Bush, Spindle Tree, Bitter Ash, Strawberry Tree, Spindle Bush, Indian Arrow Wood.
EUPATORIUM AROMATICUM	*Pool Root,* White Snake Root.
" PERFOLIATUM	*Boneset,* Thorough Root, Thorough Wax, Thoroughwort, Thorough Wax Stem, Crosswort, Thoroughwort Root, Vegetable Antimony, Indian Sage, Feverwort, Ague Weed, Sweating Plant.
" PURPUREUM..	*Queen of the Meadow,* Gravel Root, Trumpet Weed, Purple Boneset, Joe Pye.
" URTICIFOLIUM	*White Snake Root.*
EUPHORBIA COROLLATA....	*Large Flowering Spurge,* Blooming Spurge, Bowman's Root, Milk Weed.

Botanical or Officinal Names.	Local or Common Names.
EUPHORBIA ipecacuanha..	*American Ipecac*, Wild Ipecac, Ipecacuanha Spurge.
EUPHRASIA officinalis...	*Eyebright.*
FRAGARIA virginiana.........	*Strawberry Leaves.*
FRASERA verticellata........	*Columbo*, Ohio Curcuma, Golden Seal.
" walteri............ } (See *Cocculus Palmatus.*) }	*American Columbo*, Pyramid Flower, Frasera, Columbia, Marietta Columbia, Yellow Gentian, Meadow Pride, Indian Lettuce, Wild Columbo.
FRAXINUS acuminata.....	*White Ash Bark.*
FUCUS............	*Dulce.*
FUMARIA officinalis......	*Fumitory.*
GALANGA major and minor	*Catarrh Root*, East India Root, Galangal, Kassamac.
GALIPEA officinalis.......	Angustura Bark.
GALIUM aparine.................	*Cleaver's Root*, Cliver's Root, Goose Grass, Robin Run the Hedge, Milksweet, Clabber Grass, Poor Robin, Catch Weed, Bed Straw, Savoyan, Gravel Grass.
GAULTHERIA procumbens	*Wintergreen*, Partridge Berry, Spicy Wintergreen, Deer Berry, Trailing Gaultheria, Berried Tea, Grouse Berry, Ground Ivy, Checker Berry, Tea Berry, Box Berry, Spice Berry, Mountain Tea, Gaultheria, Red Berry, Ground Berry.

BOTANICAL OR OFFICINAL NAMES.	LOCAL OR COMMON NAMES.
GELSEMINUM SEMPER-VIRENS	*Yellow Jessamine,* Wild Jessamine, Carolina Jessamine, Bignonia, Woodbine.
GENTIANA CATESBÆI	*Blue Gentian,* Flux Root, Southern Gentian.
" LUTEA	*Great Yellow Gentian.*
" QUINQUEFLORA	*Bitter Plantain,* Gall of the Earth, Five-flowered Gentian.
" SAPONARIA	*Sampson Snake Root.*
GERANIUM MACULATUM	*Crane's Bill,* Stork's Bill, Spotted Crane's Bill, Tormentilla, Spotted Geranium, Crowfoot, Alum Root, Racine a Becquet, Dovefoot.
" ROBERTIANUM	*Herb Robert.*
GEUM RIVALE	*Avens Root,* Water Avens, Purple Avens, Throat Root.
" VIRGINIANUM	*White Avens,* Evan Root, Chocolate Root, Cure-all, Throat Root, Bennet.
GILLENIA STIPULACEA	*American Ipecac.*
" TRIFOLIATA	*Indian Physic,* Bowman's Root, Beaumont's Root, American Ipecac, Three-leaved Spirea, Meadow-sweet, Dropwort, Indian Hippo.

BOTANICAL OR OFFICINAL NAMES.	LOCAL OR COMMON NAMES.
GLECHOMA HEDERACEA.... (See *Nepeta Glechoma.*)	*Ground Ivy*, Gill Run, Gill go over the Ground, Cat-foot, Ale-hoof.
GLYCYRRHIZA GLABRA.......	*Liquorice*, Licorice, Sweetwood, Spanish Root.
GNAPHALIUM MARGARI- TACEUM............................ GNAPHALIUM POLYCE- PHALUM .... GNAPHALIUM ULIGINO- SUM	*Life Everlasting*, Pearly Everlasting, Cud Weed, Field Balsam, Live Forever, Sweet Balsam, White Balsam, Indian Posey, Mouse Ear, Low Cud Weed, Dysentery Root.
GOODYERA PUBESCENS.....	*Rattlesnake Root*, Adder Violet, Netleaf Plantain.
GOSSYPIUM HERBACEUM....	*Cotton Root.*
HAMAMELIS VIRGINICA....	*Witch Hazel*, Striped Alder, Winter Bloom, Snapping Hazelnut, Spotted Alder.
HEDEOMA PULEGIOIDES....	*Pennyroyal*, Squaw Mint, Tick Weed, Stinking Balm.
HELIANTHEMUM CANA- DENSE	*Frostwort*, Frost Weed, Rock Rose, Scrofula Weed.
HELONIAS DIOICA...............	*Starwort*, False Unicorn, Unicorn's Horn, Devil's-bit, Drooping Starwort, Colic Root, Blazing Star.
HELLEBORUS AMERICANA....	*Veratrum Viride*, Indian Poke, American White Hellebore, Poke Root, Swamp Hellebore, Itch Weed.

Botanical or Officinal Names.	Local or Common Names.
HELLEBORUS fœtidus........	*Bear's Foot*, Setterswort.
" NIGER............	*Black Hellebore*, Christmas Rose.
HEPATICA americana..........	*Liverwort*, Liver Leaf, Kidney Liver Leaf.
" triloba	" "
HERACLEUM lanatum........	*Masterwort*, Cow Parsnip.
HEUCHERA americana........	*Alum Root*, American Sanacle, Maple-leaf Alum Root, Cliff Weed, Ground Maple, Split Rock.
HUMULUS lupulus..............	*Hops*, Lupulus, Humulus, Lupulin.
HYDRANGEA arborescens..	*Hydrangea*, Wild Hydrangea, Seven-barks.
HYDRASTIS canadensis.......	*Golden Seal*, Yellow Root, Yellow Puccoon, Orange Root, Ground Raspberry, Indian Paint, Ohio Curcuma, Eye Balm, Turmeric Root.
HYOSCYAMUS niger............	*Black Henbane*, Fœtid Nightshade, Poison Tobacco.
HYPERICUM perforatum....	*St. Johnswort*, Johnswort.

Botanical or Officinal Names.	Local or Common Names.
HYSSOPUS OFFICINALIS.........	*Hyssop.*
IMPATIENS BALSAMINA.........	*Sweet Balsam,* Touch-me-not, Impatiens, Lady's Slipper.
" FULVA & PALLIDA	*Wild Celandine,* Balsam Weed, Jewel Weed, Touch-me-not.
IMPERATORIA OSTRUTHIUM	*Masterwort.*
INULA HELENIUM..................	*Elecampane,* Scabwort.
IPOMÆA JALAPA....................	*Jalap.*
IRIS FLORENTINA..........	*Orris Root.*
" VERSICOLOR...	*Blue Flag,* Flag Lily, Fleur-de-lis, Liver Lily, Poison Flag, Snake Lily, Water Flag.
JUGLANS CINEREA.........	*White Walnut,* Butternut, Oil Nut, Lemon Walnut.
" NIGRA..........	*Black Walnut.*
JUNIPERUS COMMUNIS.........	*Juniper.*
" SABINA................	*Savine.*
" VIRGINIANA........	*Red Cedar.*
KALMIA ANGUSTIFOLIA...........	*Sheep Laurel,* Narrow-leaved Laurel, Sheep Poison.
" GLAUCA	*Swamp Laurel.*

BOTANICAL OR OFFICINAL NAMES.	LOCAL OR COMMON NAMES.
KALMIA LATIFOLIA...............	*Laurel*, Mountain Laurel, Broad-leaved Laurel, Big-leaved Ivy, Ivy, Lambkill, Spoonwood, Calico Bush, Sheep Poison, Calfkill, Sheep Laurel.
KRAMERIA TRIANDRA......	*Rhatany.*
LACTUCA ELONGATA.......	*Wild Lettuce*, Trumpet Weed, Snake Weed, Snake Bite.
" SATIVA.........	*Garden Lettuce.*
LAPPA MINOR........... } (See *Arctium Lappa.*)	*Burdock Leaves*, Clot Burr.
LAURUS BENZOIN........ } (See *Benzoin Odoriferon.*)	*Fever Bush*, Fever Wood, Spice Bush, Spice Wood, Spice Berry, Wild Allspice, Allspice Bush.
" NOBILIS............... } (See *Myrica Cerifera.*)	*Bayberries*, Bay Tree, Laurus, Sweet Bay.
" SASSAFRAS...............	*Sassafras*, Cinnamon Wood, Saxafrax.
LAVANDULA VERA...............	*Lavender.*
LEONTICE THALICTROIDES....	*Blue Cohosh*, Pappoose Root, Blueberry.
LEONURUS CARDIACA............	*Motherwort*, Cardiaca.
LEONTODON TARAXACUM....	*Dandelion*, Monkshood.

BOTANICAL OR OFFICINAL NAMES.	LOCAL OR COMMON NAMES.
LEPTANDRA VIRGINICA.......	*Culvers Root*, Culvers Physic, Black Root, Brinton Root, Tall Veronica, Virginia, Speedwell, Veronica, Bowman's Root.
LIATRIS SPICATA..................	*Button Snake Root*, Gayfeather, Devil's-bit, Colic Root, Backache Root, Corn Snake Root, Slender-spiked Liatris.
" SQUARROSA...............	*Rattlesnake's Master*, Blazing Star.
LIGUSTICUM LEVISTICUM....	*Lovage*, Lavose, Smelage.
LIGUSTRUM VULGARE..........	*Privet*, Prim.
LIRIODENDRON TULIPIFERA	*Tulip Tree Bark*, Whitewood, White Poplar, Yellow Poplar, American Poplar, American Tulip Tree, Old Wife's Shirt, Cyprus Tree, Lyre Tree of America, Tulip-bearing Poplar, Len-nik-bi, Cucumber Tree.
LOBELIA CARDINALIS.......	*Red Cardinal Flower*, Red Cardinal Plant, Red Lobelia, Indian Pink.
" INFLATA.......	*Lobelia*, Bugle Weed, Indian Tobacco, Wild Tobacco, Asthma Root, Emetic Herb, Emetic Weed, Puke Root, Eye Bright.

BOTANICAL OR OFFICINAL NAMES.	LOCAL OR COMMON NAMES.
LOBELIA SYPHILITICA............	*Blue Lobelia*, Blue Cardinal Flower, Bladder Podded Cardinal Flower.
LYCOPODIUM CLAVATUM....	*Club Moss*, Vegetable Sulphur.
LYCOPUS EUROPÆUS.......	*Green Archangel*, Water Hoarhound, Water Bugle, Gipseywort.
" VIRGINICUS..............	*American Archangel*, Red Archangel, Gipsey Weed, Virginia Hoarhound, Paul's Betony, Sweet Bugle, Bugle Weed, Bitter Bugle, Water Bugle.
LYTHRUM SALICARIA............	*Loose Strife*, Spiked Strife, Purple Willow Herb.
MACROTYS RACEMOSA............	*Black Snake Root*, Black Cohosh, Squaw Root, Rattle Root, Rich Weed.
MAGNOLIA GLAUCA........	*Sweet Magnolia*, Beaver Tree, Swamp Laurel, Sweet Bay, White Bay, Swamp Sassafras.
MALVA ROTUNDIFOLIA......	*Low Mallows*, Cheese Plant.
" SYLVESTRIS........ .	*°Common Mallows*, High Cheese Plant.
MARRUBIUM VULGARE.....	*Prassium*, White Hoarhound.
MATRICARIA CHAMOMILLA..	*German Chamomile.*
MEDEOLA VIRGINICA.......	*Indian Cucumber*, Cucumber Root.

BOTANICAL OR OFFICINAL NAMES.	LOCAL OR COMMON NAMES.
MELIA AZEDARACH...............	*Pride of China,* Pride of India, Bead Tree.
MELILOTUS OFFICINALIS......	*Melilot,* King's Clover, Sweet Clover, Yellow Melilot.
MELISSA "	*Lemon Balm,* Melissa, Dropsy Plant, Cure-all, King's Clover, Sweet Balm.
MENISPERMUM CANADENSE	*Texas Sarsaparilla,* Yellow Parilla, Canada Wormwood, Canadian Moonseed, Vine Maple.
MENTHA PIPERITA.................	*Peppermint.*
" VIRIDIS....................	*Spearmint.*
MENYANTHES TRIFOLIATA..	*Buck Bean,* Bog Bean, Marsh Trifoil, Water Shamrock, Bitter-worm.
MEZEREUM DAPHNE........	*Mezereon.*
MIMULUS MOSCHATUS......	*Musk Plant,* Monkey Flower.
MITCHELLA REPENS..............	*Squaw Vine,* Winter Clover, One Berry, Partridge-berry Vine, Hive Vine.
MOMORDICA BALSAMINA......	*Balsam Apple,* Partridge Berry, Balsamine.
MONARDA DIDYMA...............	*Mountain Balm,* Mountain Mint, Oswego Tea, Square, Red Balm, Square Stalk.

BOTANICAL OR OFFICINAL NAMES.	LOCAL OR COMMON NAMES.
MONARDA PUNCTATA............	*Horse Mint.*
MONOTROPA UNIFLORA..,......	*Fit Root,* Fit Plant, Pipe Plant, Bird Nest, Pine Sap, Corpse Plant, Ice Plant, Indian Pipe, Ova-ova.
MYRICA CERIFERA.................	*Bayberry,* Candleberry, Wax-berry, Wax Myrtle, Myrtle Bayberry Tree.
" GALE......................	*Sweet Gale,* Sweet Willow, Dutch Myrtle, Bog Myrtle, Meadow Fern, Bay Bush.
NABULA ALBUS...................	*Cancer Weed,* Lion's Foot, White Lettuce, Rattlesnake Root.
NARCISSUS PSEUDO-NAR-CISSUS .....:..................... }	*Narcissus,* Daffodil.
NARD........ ....} (See *Aralia Racemosa.*) }	*Spikenard,* Nardus, Life of Man, Petty Morel, Spignet.
NASTURTIUM AMERICANA } (See *Tropæolum Majus.*) }	*Water Radish,* Water Cress, Amphibious Cress.
NASTURTIUM AMPHIBIUM...	" "
" PALUSTRE.....	*Marsh Water Cress.*
NEPETA CATARIA..................	*Catnip,* Cat Mint, Nepeta Cat wort.
" GLECHOMA	See *Glechoma Hederacea.*

BOTANICAL OR OFFICINAL NAMES.	LOCAL OR COMMON NAMES.
NUPHAR ADVENA NYMPHÆA ODORATA	*Yellow Pond Lily,* Toad Lily, Frog Lily, Cow Lily, Water Lily, Water Nymph, Cow Cabbage, White Lily, White Pond Lily, Sweet Water Lily, Water Cabbage, Spatter Dock, Beaver Root.
OCYMUM BASILICUM	*Sweet Basil,* Royal Ocimum, Basilicum.
ŒNANTHE CROCATA	*Hemlock Water Drop.*
" PHELLANDRIUM	*Fine-leaved Water Hemlock.*
ŒNOTHERA BIENNIS	*Evening Primrose,* Tree Primrose, Scabish.
ORIGANUM MAJORANA	*Sweet Marjoram,* Sampscus Marjoram, Amaracus.
" VULGARE	*Mountain Mint,* Wild Marjoram.
OROBANCHE VIRGINIANA " UNIFLORA	*Cancer Root,* Beech Drop, Broom Rape, Squaw Root.
OSMORRHIZA BREVISTYLIS	*Sweet Cicily,* Short-styled Cicily, Hairy Sweet Cicily.
OXALIS ACETOSELLA	*Wood Sorrel,* Acetosella, Cuckoo's Bread.
PÆONIA OFFICINALIS	*Peony.*

BOTANICAL OR OFFICINAL NAMES.	LOCAL OR COMMON NAMES.
PANAX QUINQUEFOLIA............	*Ginseng*, Ninsin, Garantogen, Gensang, Five Fingers, Red Berry.
PAPAVER SOMNIFERUM..........	*Poppy*, Opium Poppy.
PARTHENIUM INTEGRI- FOLIUM .... }	*Prairie Dock*, Nephritic Plant, Cutting Almond.
PEDICULARIS CANADENSIS...	*Wood Betony*, Betony Weed, Lousewort.
PERSICA, AMYGDALUS............	*Peach Leaves*, Nectarine.
PETROSELINUM SATIVUM..	*Parsley Root*, Rock Parsley.
PHYTOLACCA DECANDRA.....	*Poke*, Garget, Cocum, Pigeon Berry, Jalap, American Nightshade, Jalap Cancer Root, Pecatacalleloe, Skoke Root, Scoka.
PIMPINELLA, SAXIFRAGA....	*Small Burnet Saxifrage*, Saxifrage.
PINCKNEYA PUBENS............	*Pinckneya*, Carolina Cinchona, Georgia Bark, Fever Tree, Florida Bark.
PINUS CANADENSIS................	*Hemlock Spruce.*
" BALSAMEA	*Balsam Fir.*
PIPER ANGUSTIFOLIUM............	*Matico*, Narrow-leaved Piper.

BOTANICAL OR OFFICINAL NAMES.	LOCAL OR COMMON NAMES.
PLANTAGO MAJOR...............	*Plantain*, Greater Plantain, Way Bread, Rib Grass.
PODOPHYLLUM PELTATUM	*May Apple*, Mandrake, Wild Lemon, Indian Apple, Wild Mandrake, Raccoon Berry, Duck's Foot, Ipecacuanha.
POLEMONIUM REPTANS......	*Abscess Root*, Blue Bell, Jacob's Ladder, Sweat Root, Greek Valerian, American Valerian.
POLYGALA RUBELLA..............	*Bitter Polygala*, Ground Flower.
" SENEGA..............	*Seneka Snake Root*, Seneka, Senega, Rattlesnake Root, Mountain Flax, Milkwort, Snake Weed.
POLYGONUM HYDROPIPER...	*Smart Weed*, Arsmart, Water Pepper.
" PUNCTATUM ...	*Water Pepper*, Dead Arsmart, Knot Weed, Biting Persicaria, Biting Knot Weed.
POLYPODIUM VULGARE.......	*Common Polypody*, Brake Root, Rock Polypody.
POLYTRICHUM JUNIPE- RINUM......	*Hair-cap Moss*, Robin's Rye, Bear's Bed, Ground Moss.
POPULUS BALSAMIFERA.....	*Balsam Poplar*, Tacamahac, Balm of Gilead.

BOTANICAL OR OFFICINAL NAMES.	LOCAL OR COMMON NAMES.
POPULUS TREMULOIDES.........	*Trembling Poplar, Aspen*, American Poplar, White Poplar, American Aspen, Quaking Aspen, Quiver Leaf.
POTENTILLA CANADENSIS....	*Cinque Foil*, Five Finger.
PRINOS VERTICILLATUS..........	*Black Alder*, False Alder, Striped Alder, Winter-berry, Virginia Winter-berry, Fever Bush.
PRUNELLA PENNSYLVANICA..	*Heal-all*, Self-heal.
PRUNUS VIRGINIANA..............	*Wild Cherry Bark*, Black Cherry, Cabinet Cherry, Rum Cherry.
PTELEA TRIFOLIATA...............	*Wafer Ash*, Shrubby Trefoil, Hop Tree, Ptelea Bark, Ague Bark, Wing Seed, Winter Fern.
PTERIS ATROPURPUREA.....	*Rock Brake*, Winter Fern, Indian Dream.
PULMONARIA OFFICINALIS..	*Lungwort*, Cowslips of Jerusalem, Gum Plant, Healing Herb, Jacob's Ladder, Sage of Jerusalem, Spotted Comfrey, Spotted Lungwort, Maple Lungwort.
PYRETHRUM PARTHENIUM..	*Feverfew*, Featherfew, Chrysanthemum.

BOTANICAL OR OFFICINAL NAMES.	LOCAL OR COMMON NAMES.
PYROLA MACULATA........	*King's Cure.*
" ROTUNDIFOLIA.....	*Wild Lettuce,* Round-leafed Pyrola.
" UMBELLATA.............	*Round-leafed Consumption Weed,* Shin Leaf, Ground Holly, King's Cure.
QUERCUS ALBA....................	*White Oak Bark.*
" RUBRA...................	*Red Oak Bark.*
" TINCTORIA	*Black Oak Bark.*
QUILLAYA SAPONARIA......	*Soap Tree.*
RANUNCULUS ACRIS.. ...	*Buttercup,* Meadow Crowfoot, Yellow Weed, Meadow Bloom, Blister Weed.
" BULBOSUS...	*Crowfoot,* Bulbous Crowfoot.
RHAMNUS CATHARTICUS....	*Buck Thorn,* Purging Berries.
RHODODENDRON CHRY-} SANTHEMUM..........)	*Yellow-flowered Rhododendrum.*
RHUS GLABRA.............	*Smooth Sumach,* Pennsylvania Sumach, Upland Sumach.
" TOXICODENDRON............	*Poison Oak,* Poison Ivy, Poison Ash, Poison Vine.
ROSMARINUS OFFICINALIS....	*Rosemary,* Rosemarinus.
RUBUS OCCIDENTALIS.............	*Blackberry,* Thimbleberry, Black Raspberry.

BOTANICAL OR OFFICINAL NAMES.	LOCAL OR COMMON NAMES.
RUBUS STRIGOSUS	*Raspberry Leaves*, Wild Red Raspberry.
" VILLOSUS	*Blackberry Root*, High Blackberry, Thimbleberry, Hairy American Bramble.
" TRIVIALIS	*Dewberry*, Creeping Blackberry.
RUDBECKIA LACINIATA	*Thimble Weed*, Cone-disk Sun Flower, Tall Cone Flower.
RUMEX AQUATICUS	*Water Dock*, Xanthoriza, Hydrola Pathrum, Great Yellow Dock.
" CRISPUS	*Yellow Dock*, Curled Dock, Narrow Dock, Sour Dock, Garden Patience.
RUTA GRAVEOLENS	*Rue*, Garden Rue.
SABBATIA ANGULARIS	*Red Century*, American Century, Sabbatia, Angular-stalked Sabbatia, Angular-stalked Star Flower, Bitter Bloom, Bitter Clover, Rose Pink, Wild Succory.
SALIX ALBA	*White Willow*, Willow Bark.
" NIGRA	*Pussy Willow*.
SALVIA OFFICINALIS	*Sage*, Garden Sage.
" LYRATA	*Wild Sage*, Meadow Sage, Lyre-leaved Sage, Cancer Weed.

BOTANICAL OR OFFICINAL NAMES·	LOCAL OR COMMON NAMES.
SALVIA SCLARY......................	*Clarea,* Sclarea, Clary, Clamy Sage.
SAMBUCUS CANADENSIS........	*Elder Blooms,* Common Elder, Sweet Elder.
SANGUINARIA "	*Blood Root,* Red Root, Puccoon, Tetterwort, Indian Paint.
SANICULA MARILANDICA....	*Sanicle,* Black Snake Root, Pool Root.
SAPONARIA OFFICINALIS....	*Soapwort,* Bouncing Bet, Old Maid's Pink, London Pride.
SARRACENIA PURPUREA.. } " OFFICINALIS }	*Side-saddle Plant,* Fly-trap, Water Cup, Pitcher Plant, Eve's Cup, Huntsman's Cup, Small Pox Plant.
SASSAFRAS OFFICINALIS........	*Sassafras Bark.*
SATUREIA HORTENSIS...........	*Summer Savory.*
SCOPARIUS CYTISUS..............	*Broom.*
SCROPHULARIA NODOSA....	*Figwort,* Scrofula Plant, Square Stalk, Knotty-rooted Figwort, Carpenter's Square, Heal-all.
SCUTELLARIA LATERIFLORA	*Scullcap,* Hooded Willow Herb, Hoodwort, Mad Weed, Mad Dog Scullcap, Blue Scullcap, Blue Pimpernel,

BOTANICAL OR OFFICINAL NAMES.	LOCAL OR COMMON NAMES.
SENECIO AUREUS....................	*Life Root*, Golden Ragwort, Uncum, Squaw Weed, Golden Senecio.
" GRACILIS..........	*Female Regulator.*
" HIERACIFOLIUS........	*Fire Weed.* (See *Erecthites Hieracifolius.*)
" OBOVATUS...............	*Squaw Weed.*
" VULGARIS	*Common Groundsel.*
SESAMUM FOLIA....................	*Benne Leaves.*
SILENE VIRGINICA...................	*Catch-fly*, Wild Pink, Virginia Catch-fly, Ground Pink.
SIMARUBA EXCELSA...............	*Quassia*, Bitter Wood, Bitterwort, Simaruba, Mountain Zarusum, Bitter Weed.
SISYMBRIUM OFFICINALE.....	*Hedge Mustard.*
SIUM NODIFLORUM..................	*Water Parsnip.*
SMILAX OFFICINALIS...............	*Sarsaparilla*, Wild Liquorice.
" HERBACEA..................	*Carrion Flower.*
" PEDUNCULARIS...........	*Jacob's Ladder.*
SOLANUM DULCAMARA....'......	*Bittersweet*, Woody Nightshade, Bittersweet Nightshade, Garden Nightshade, Dulcamara, Scarlet Berry, Violet Bloom.

BOTANICAL OR OFFICINAL NAMES.	LOCAL OR COMMON NAMES.
SOLIDAGO ODORA.................	*Golden Rod,* Sweet-scented Golden Rod.
SPIGELIA MARILANDICA..........	*Carolina Pink Root,* Wormgrass, Wormseed, Indian Pink Root, Unsteetle, Star Bloom.
SPIRÆA TOMENTOSA...............	*Hardhack,* Steeplebush, Whitecap, Meadow Sweet, White Leaf.
" ULMARIA...........	*Queen of the Meadow.*
STATICE CAROLINIANA......	*Marsh Rosemary,* Seathrift, Meadow Root, Ink Root, Sea Lavender.
STILLINGIA SYLVATICA.....	*Queen's Root,* Queen's Delight, Cock-up-hat, Stillingia, Yaw Root, Silver Leaf.
SYMPHYTUM OFFICINALE..	*Comfrey,* Healing Herb, Gum Plant.
SYMPLOCARPUS FŒTIDUS..	See *Dracontium Fœtidum.*
TANACETUM VULGARE.....	*Double Tansy,* Tansy.
TEPHROSIA VIRGINIANA....	*Turkey Pea,* Turkey Corn, Goat's Rue, Hoary Pea, Devil's Shoestring, Catgut.
THLASPI BURSI PASTORIS...	*Shepherd's Purse.*
THUJA OCCIDENTALIS.......	*Arbor Vitæ,* False White Cedar.
THYMUS VULGARIS.........	*Thyme,* Garden Thyme, Thymus, Mother of Thyme.

BOTANICAL OR OFFICINAL NAMES.	LOCAL OR COMMON NAMES.
TRIFOLIUM PRATENSE.........	*Red Clover.*
" REPENS	*White Clover.*
TRILLIUM PENDULUM...........	*Beth Root,* Birth Root, Cough Root, Ground Lily, Wake Robin, Indian Balm, Jews-harp, Lamb's Quarter, Snake-bite, Pariswort, True Love.
TRIOSTEUM PERFOLIATUM..	*Fever Root,* Feverwort, Wild Ipe-cac, Bastard Ipecac, Doctor Tinker's Weed, False Ipecac, Cinque, Horse Ginseng, White Gentian, Witch Grass, Quickens, Dog Grass, Wild Coffee, Sweet Bitter.
TRITICUM REPENS.........	*Couch Grass,* Quich Grass, Quitch Grass, Dog Grass, Knot Grass, Witch Grass, Quickens.
TROPÆOLUM MAJUS.......	*Nasturtiun,* Indian Cress.
TUSSILAGO FARFARA......	*Coltsfoot,* Bull's Foot, Flower Velure.
ULMUS FULVA............	*Slippery Elm,* Indian Elm, Sweet Elm, Red Elm.
URTICA DIOICA............	*Nettle Root,* Stinging Nettle.
VALERIANA OFFICINALIS...	*Valerian,* Great Wild Valerian.
VARIOLARIA FAGINICA.....	*Lady's Slipper.*
VERATRUM ALBUM........	*White Hellebore.*

BOTANICAL OR OFFICINAL NAMES.	LOCAL OR COMMON NAMES.
VERATRUM VIRIDE.............	*American Hellebore*, Itch Weed, Indian Poke, Swamp Hellebore, Wolfbane.
VERBASCUM THAPSUS..........	*Mullein*, Yellow Moth, Itch Weed, Blattaria.
VERBENA HASTATA..............	*Blue Vervain*, Purvain, Simpler's Joy, Wild Hyssop.
" OFFICINALIS..........	*Vervain*, Verbena, Simpler's Joy.
" URTICIFOLIA.........	*White* or Nettle-leaved Vervain.
VERONICA OFFICINALIS.........	*Speedwell*, Fluellin, Virginia Speedwell.
VIBURNUM OPULUS..............	*Cramp Bark*, High Cranberry, Snow Ball, Cranberry Tree, Nanny Bush Bark, Guelder's Rose, Sheep's Berry.
" PRUNIFOLIUM..	*Black Haw*, Sloe.
VIOLA CUCULLATA..........	*Blue Violet.*
" PEDATA.............	*Birdsfoot Violet.*
" ROSTRATA..........	*Canker Violet.*
XANTHEUM STRUMARIUM...	*Cockle Bur*, Clot Bur.
ZANTHORIZA APIIFOLIA....	*Yellow Root*, Shrub Yellow Root.
ZANTHOXYLUM FRAXINEUM	*Yellow Wood*, Prickly Ash, Suterberry, Pellitory, Toothache bush, Toothache Tree, Angelica Tree Bark, Parsley Yellow Root.
ZEDOARY RADIX........,.....	*Zedoary Root.*
ZINGIBER OFFICINALIS.......	*Ginger.*

ERRATA.

Put Aletris farinosa in italic, p. 44.

Nabulus Albus is incorrectly spelt on pp. 29, 90, 106, 107.

Change the *Officinal* of Zanthoriza to Zanthoriza apiefolia, p. 137.

"   of Parsley Yellow Root to   "

PART II.

Local or Common Names.	Botanical or Officinal Names.
Abscess Root...................	Polemonium reptans.
Acetosella......................	Oxalis acetosella.
Aconite........................	Aconitum·napellus.
Adam and Eve..................	Corallorhiza hyemalis.
Adder's Violet...........	Goodyera pubescens.
Agrimony...............	Agrimonia eupatoria.
Ague Bark..............	Ptelea trifoliata.
" Grass.............	Aletris farinosa.
" Root................	Aletris farinosa.
" Weed...............	Eupatorium perfoliatum.
Alder, Black...................	Prinos verticillatus.
" False...................	" "
" Red.....................	Alnus rubra.
" Smooth.................	"
" Striped................	*Hamamelis virginica.*
" "	Prinos verticillatus
" Swamp.................	Alnus rubra.

LOCAL OR COMMON NAMES.	BOTANICAL OR OFFICINAL NAMES.
Alder, Tag	Alnus rubra.
Alehoof	Nepeta glechoma.
Alkanet Root	Anchusa Tinctoria.
Aloe Root	Aletris farinosa.
" "	Agave virginica.
Allspice Bush	Laurus benzoin.
Althea	Althæa officinalis.
Alum Root	*Heuchera Americana.*
" "	Geranium maculatum.
Amaracus	Origanum majorana.
Amaranth	Amaranthus hypochondriacus.
American Archangel	Lycopus Virginicus.
" Aspen	Populus tremuloides.
" Century	Centaurea Americana.
" Columbo	Frasera walteri.
" Dittany	Cunila mariana.
" Hellebore	Veratrum viride.
" Ipecac	*Euphorbia ipecacuanha.*

LOCAL OR COMMON NAMES.	BOTANICAL OR OFFICINAL NAMES.
American Ipecac.............	Gillenia trifoliata.
·· **Nightshade**	Phytolacca decandra.
" **Poplar**..............	Liriodendron tulipifera.
" **Red Cornel**......	Cornus sericea.
" **Saffron**.............	Carthamus tinctorius.
" **Sanicle**..............	*Sanicula marilandica.*
" "	Heuchera Americana.
" **Senna**...............	Cassia marilandica.
" **Tulip Tree**........	Liriodendron tulipifera.
" **Valerian**...........	*Cypripedium pubescens.*
" "	Polemonium reptans.
" **Wintergreen**...	Chimaphila umbellata.
Amphibious Water Cress	*Nasturtium amphibium.*
" " "	" Americana.
Anemone, Meadow...........	Anemone pratensis.
Angelica leaves or bark.	Angelica atropurpurea.
Angelica Tree bark...........	*Aralia spinosa.*

LOCAL OR COMMON NAMES.	BOTANICAL OR OFFICINAL NAMES.
Angelica Tree Bark.....	Zanthoxylum fraxineum.
" Wild..........	Angelica sylvestris.
Angular Stalked Sabbatia	Sabbatia angularis.
" Stemmed Star-flower..........	" "
Angustura Bark...............	Galipea officinalis.
Antidote............................	Dorstenia contrayerva.
Apple, Balsam.....................	Momordica balsamina.
" Devil's....................	Datura stramonium.
" May.....................	Podophyllum peltatum.
" of Peru..................	Datura stramonium.
" Thorn.....................	" "
Arbor Vitæ.......................	Thuja occidentalis.
Arbutus, Trailing.......	Epigæa repens.
Archangel, Wild........	Angelica archangelica.
Arnica Blooms...................	Arnica montana.
Arsmart	Polygonum hydropiper.
Asclepias, Flesh-colored.	Asclepias incarnata.
Ash, Prickly......................	*Zanthoxylum fraxineum.*

LOCAL OR COMMON NAMES.	BOTANICAL OR OFFICINAL NAMES.
Ash, Prickly	Aralia spinosa.
" Wafer	Ptelea trifoliata.
" White	Fraxinus acuminata.
Aspen, Trembling	Populus tremuloides.
Asthma Root	Lobelia inflata.
Australian Blue Gum	Eucalyptus globulus.
Avens, Purple	Geum rivale
" Root	" "
" Water	" "
" White	" "
Azedarach	Melia azedarach.
Backache-brake	Asplenium filix fœmina.
" " Root	" "
" Root	Liatris spicata.
Balm, Horse	Collinsonia canandensis.
" Lemon	Melissa officinalis.
" Mountain	Monarda didyma.
Balm, Red	Monarda didyma.

LOCAL OR COMMON NAMES.	BOTANICAL OR OFFICINAL NAMES.
Balm of Gilead.........	Populus balsamifera.
" " Herb.....	Dracocephalum canariense.
" Sweet........... ..	*Melissa officinalis.*
" "	Dracocephalum canariense.
Balmony........................	Chelone glabra.
Balsam Apple..................	Momordica balsamina.
" Fir....................	Pinus balsamea.
" Poplar................	Populus balsamifera.
" Sweet.................	*Gnaphalium polycephalum.*
" "	Impatiens balsamina.
" Weed..........	" fulva and palida.
" White..........	{ *Gnaphalium polycephalum.* Impatiens fulva and palida.
Balsamina........................	Momordica balsamina.
Bane Berry....................	Actea alba and rubra.
Barberry	Berberis vulgaris.
Basilicum	Ocymum basilicum.
Bastard Ipecac................	Triosteum perfoliatum.
" Saffron................	Carthamus tinctorius.
Bayberry........................	Myrica cerifera.
" Bush................	Myrica gale.

LOCAL OR COMMON NAMES.	BOTANICAL OR OFFICINAL NAMES.
Bay Tree	*Myrica cerifera.*
" "	Laurus nobilis.
Bead Tree	Melia azedarach.
Bearberry	Arctostaphylos uva ursi.
Bear's Bed	Polytrichum juniperinum.
Bear's Foot	Dracontium fœtidum.
"	*Helleborus fœtidus.*
Bear's Leaf	Dracontium fœtidum.
Bear's Whortleberry	Arctostaphylos uva ursi.
Beaumont's Root	Gillenia trifoliata.
Beaver Poison	Conium maculatum.
" Root	Nuphar advena.
" Tree	Magnolia glauca.
Bedstraw	Galium aparine.
Beech Drop	*Orobanche Virginiana.*
"	Epiphegus Americanus.
Beerenwortel	Dracontium fœtidum.
Bee's Nest	Daucus carota.
Belladonna	Atropa belladonna.

LOCAL OR COMMON NAMES.	BOTANICAL OR OFFICINAL NAMES.
Benjamin Bush	Benzoin odoriferum.
Benne Leaves	Sesamum folia.
Bennet	Geum Virginianum.
Berried Tea	Gaultheria procumbens.
Bethroot	Trillium pendulum.
Betony Weed	Pedicularis canadensis.
" Wood	" "
Big-leaved Ivy	Kalmia latifolia.
Bignonia	Gelseminum sempervirens.
Bindweed	Convolvulus panduratus.
Bion-retter	Dracontium fœtidum.
Biorn-blad	" "
Birch Bark	Betula lenta.
Birdsfoot Violet	Viola pedata.
Bird's Nest	*Daucus carota.*
"	Monotropa uniflora.
Bird Pepper	Capsicum annuum.
Birth Root	Trillium pendulum.

LOCAL OR COMMON NAMES	BOTANICAL OR OFFICINAL NAMES.
Birthwort	Aristolochia serpentaria.
Bismalva	Althæa officinalis.
Biting Knotweed	Polygonum punctatum.
Biting Persicaria	" "
Bitter Ash	Euonymus atropurpureus.
" **Apple**	Cucumis colocynthis.
" **Bloom**	Sabbatia angularis.
" **Bugle**	Lycopus Virginicus.
" **Clover**	*Sabbatia angularis.*
" "	Centaurea Americana.
" **Cucumber**	Cucumis colocynthis.
" **Dogsbane**	Apocynum androsæmifolium.
" **Grass**	Aletris farinosa.
" **Herb**	Chelone glabra.
" **Plantain**	Gentiana quinqueflora.
" **Polygala**	Polygala rubella.
" **Root**	Apocynum androsæmifolium.
" **Stick**	Agathotes chirayta.

LOCAL OR COMMON NAMES.	BOTANICAL OR OFFICINAL NAMES.
Bitter Weed...............	Simaruba excelsa.
" **Wood**...............	" "
" "	*Quassia excelsa.*
" **Worm**...............	Menyanthes trifoliata.
" **Wort**...............	Simaruba excelsa.
Bittersweet Bark........	Celastrus scandens.
" **Climbing**...	" "
" **Herb**..............	Solanum dulcamara.
" **Nightshade**..	" "
" **Weed**.............	Apocynum androsæmifolium.
Black Alder...................	Prinos verticillatus.
" **Berry**...............	Rubus occidentalis.
" " **Root**...............	" villosus.
" **Cherry**...............	Prunus Virginiana.
" **Cohosh**...............	Cimicifuga racemosa.
" "	Macrotys "
" **Haw**..............	Viburnum prunifolium.
" **Hellebore**..........	Helleborus niger.
" **Hemp**..............	Apocynum cannabinum.

LOCAL OR COMMON NAMES.	BOTANICAL OR OFFICINAL NAMES.
Black Henbane	Hyoscyamus niger.
" Oak Bark	Quercus tinctoria.
" Raspberry	Rubus occidentalis.
" Root	*Leptandra Virginica.*
" "	Aletris farinosa.
" Snakeroot	Cimicifuga racemosa.
" "	Macrotys "
" Walnut	Juglans nigra.
Bladder-podded Lobelia	Lobelia syphilitica.
Blattaria	Verbascum thapsus.
Blazing Star	*Aletris farinosa.*
" "	Helonias dioica.
" "	Liatris squarrosa.
Bleeding Heart	Cypripedium pubescens.
Blessed Thistle	Centaurea benedicta.
Blister Weed	Ranunculus acris.
Blood Root	Sanguinaria canadensis.
Blood-stanch	Erigeron canadense.
Bloom Shell Flower	Chelone glabra.
Blue Balm	Melissa officinalis.
" Bell	Polemonium reptans.

LOCAL OR COMMON NAMES.	BOTANICAL OR OFFICINAL NAMES.
Blue Berried Cornus........	Cornus sericea.
" " Dogwood...	" "
" Berry....................	Leontice thalictroides.
" " Root................	Caulophyllum thalictroides.
" Cardinal Flower......	Lobelia syphilitica.
" Cohosh	*Caulophyllum thalictroides.*
" "	Leontice thalictroides.
" Flag....................	Iris versicolor.
" Gentian................	Gentiana catesbæi.
" Lobelia................	Lobelia syphilitica.
" Pimpernel..............	Scutellaria lateriflora.
" Scullcap..............	" "
" Violet................	Viola cucullata.
Bog Bean....................	Menyanthes trifoliata.
" Onion................	Arum triphyllum.
" Myrtle................	Myrica gale.
Bohea...	Ceanothus americana.
Boneset.................	Eupatorium perfoliatum.

LOCAL OR COMMON NAMES.	BOTANICAL OR OFFICINAL NAMES.
Bonsem Krenia	Dracontium fœtidum.
Borage	Borago officinalis.
Bouncing Bet	Saponaria "
Bowman's Root	*Gillenia trifoliata.*
" "	Euphorbia corollata.
" "	Leptandra Virginica.
Box Berry	Gaultheria procumbens.
" Tree	Cornus Florida.
" Dogtree	" "
" Wood	" "
Boy's Love	Artemisia abrotanum.
Brake-root	Polypodium vulgare.
Branching Larkspur	Delphinium consolidum.
Brinton Root	Leptandra Virginica.
Bristlestem Sarsparilla	Aralia hispida.
Bristly Aralia	" "
Broad-leaved Laurel	Kalmia latifolia.

LOCAL OR COMMON NAMES.	BOTANICAL OR OFFICINAL NAMES.
Broom	Baptisia tinctoria.
" Rape	Epiphegus Americanus.
" "	*Orobanche Virginiana.*
Buchu	Barosma crenata.
Buck Bean	Menyanthes trifoliata.
" Thorn	Rhamnus catharticus.
Buckeye	Æsculus glabra.
Budwood	Cornus Florida.
Bugbane	Cimicifuga racemosa.
Bugle	Ajuga reptans.
" Sweet	Lycopus Virginicus.
" Weed	*Lycopus Virginicus.*
" "	Lobelia inflata.
Bugloss	*Anchusa tinctoria.*
"	Borago officinalis.
Bulbous Crowfoot	Ranunculus bulbosus.
Bull's Foot	Tussilago farfara.
Burdock	Lappa minor.
" Root and Leaves	Arctium lappa.
Burning Bush	Euonymus atropurpureus.

LOCAL OR COMMON NAMES.	BOTANICAL OR OFFICINAL NAMES.
Burrage	Borago officinalis.
Butter Cup	Ranunculus acris.
Butterfly-weed	Asclepias tuberosa.
Butter Horse-weed	Erigeron canadense.
Butternut	Juglans cinerea.
Butterweed	Erigeron canadense.
Button Snakeroot	*Liatris spicata.*
" "	Eryngium aquaticum.
Cabinet Cherry	Prunus Virginiana.
Calamus	Calamus aromaticus.
Calandine	Chelidonium majus.
" Wild	Impatiens fulva and pallida.
Calf kill	Kalmia latifolia.
Calico Bush	" "
Canada Fleabane	Erigeron canadense.
" Ginger	Asarum canadense.
" Moonseed	Menispermum canadense.
" Snakeroot	Asarum canadense.

LOCAL OR COMMON NAMES.	BOTANICAL OR OFFICIAL NAMES
Canada Root............	Asclepias tuberosa.
" **Turnip**..........	Arum triphyllum.
" **Wormwood**.....	Menispermum canadense.
Canadian Wormwood...	Collinsonia canadensis.
Cancer Root.............	*Orobanche Virginiana.*
" "	Epiphegus Americanus.
" "	Phytolacca decandra.
" **Weed**	*Salvia lyrata.*
" "	Nabula albus.
Candleberry............	Myrica cerifera.
Canella, White..........	Canella alba.
Canker Root............	*Coptis trifolia.*
" "	Salvia lyrata.
" **Violet**..........	Viola rostata.
Cannabis.................	Cannabis indica.
Cardiaca.................	Leonurus cardiaca.
Cardinal Flower, Red	Lobelia cardinalis.
" **Plant,** "	" "
" **Flower, Blue**.....	" syphilitica.
Cardus Plant................	Centaurea benedicta.
" **Spotted**..................	" "

LOCAL OR COMMON NAMES.	BOTANICAL OR OFFICINAL NAMES.
Carolina Cinchona............	Pinckneya pubens.
" Jessamine..........	Gelseminum sempervirens.
" Pink Root..........	Spigelia marilandica.
Carpenter's Square.......	Scrophularia nodosa.
Carrion Flower.................	Smilax herbacea.
Carrot, Garden..................	Daucus carota.
" Wild...............`...	" "
Cascarilla......................	Croton eleuteria.
Catarrh Root....................	Galanga major and minor.
Catch Weed....................	Galium aparine.
" Fly.................	*Apocynum androsæmifolium.*
" "	Silene Virginica.
Catfoot.......................	Nepeta glechoma.
Catgut........................	Tephrosia Virginiana.
Catmint.......................	Nepeta cataria.
Catnip........................	" "
Catwort.......................	" "

LOCAL OR COMMON NAMES.	BOTANICAL OR OFFICINAL NAMES.
Cayenne Pepper	Capsicum annuum.
Cedar, Red	Juniperus Virginiana.
Celandine	Chelidonium majus.
Century, Common European	Eyrthræa centaurium.
Century, Great American	" Americana.
" Red	Sabbatia angularis.
Chamomile, German	Matricaria chamomilla.
" Low	Anthemis nobilis.
" Roman	" "
" Wild	" cotula.
Checker Berry	*Gaultheria procumbens.*
" "	Arctostaphylos uva ursi.
" "	Mitchella repens.
Cheese Plant	Malva rotundifolia.
Cherry Bark, Wild	Prunus Virginiana.
Chicken's Toes	Corallorhiza odontorhiza.
China Root	Dioscorea villosa.

LOCAL OR COMMON NAMES.	BOTANICAL OR OFFICINAL NAMES.
Chirayita	Agathotes chirayita.
Chirayit	" "
Chiretta	" "
Chocolate Root	Geum Virginiana.
Christmas Rose	Helleborus niger.
Chrysanthemum	Pyrethrum parthenium.
Cicily, Sweet	Osmorrhiza brevistylis.
Cicuta	Conium maculatum.
Cinnamon, Wild	Canella alba.
" Wood	Laurus sassafras.
Cinchona	Cinchona.
" Caroliniana	Pinckneya pubens.
Cinque	Triosteum perfoliatum.
" Foil	Potentilla canadensis.
Clabber Grass	Galium aparine.
Clammy Sage	Salvia sclary.
Clarea	" "

Local or Common Names.	Botanical or Officinal Names.
Clary	Salvia sclary.
Cleaver's Root	Galium aparine.
Cliff Weed	Heuchera Americana.
Climbing Bittersweet	Celastrus scandens.
" Staff Tree	" "
Cliver's Root	Galium aparine.
Clot Burr	*Lappa minor.*
" "	Xanthium strumarium.
" "	Arctium lappa.
Clover, Bitter	Centaurea Americana.
" King's	*Melilotus alba.*
" "	Melissa officinalis.
" Red	Trifolium pratense.
Club Moss	Lycopodium clavatum.
Cocash	Erigeron purpureum.
" Root	Aster puniceus.
Cockle Burr	Xanthium strumarium.
Cockspur Pepper	Capsicum annuum.

Local or Common Names.	Botanical or Officinal Names.
Cock-up-Hat	Stillingia sylvatica.
Cocum	Phytolacca decandra.
Cohosh, Black	Cimicifuga racemosa.
" "	Macrotys "
" **Blue**	*Caulophyllum thalictroides.*
" "	Leontice thalictroides.
" **Red**	Actea rubra.
" **White**	" alba.
Colchicum	Colchicum autumnale.
Cold Water Root	Aster puniceus.
Colic Root	*Aletris farinosa.*
" "	Asclepias tuberosa.
" "	Dioscorea villosa.
" "	Helonias dioica.
" "	Liatris spicata.
Collard	*Dracontium fœtidum.*
"	Symplocarpus fœtidus.

LOCAL OR COMMON NAMES.	BOTANICAL OR OFFICINAL NAMES.
Colocynth............................	Cucumis colocynthis.
Coltsfoot..............................	*Tussilago farfara.*
"	Asarum canadense.
" False......................	" "
" Snake Root.........	" "
Coltstail...................................	Erigeron "
Columba...............................	Cocculus palmatus.
Columbia...............................	" "
"	*Frasera walteri.*
" American..........	" "
" Marietta.............	Cocculus palmatus.
Columbo............................	*Frasera walteri.*
"	" verticillata.
Comfrey..............................	Symphitum officinale.
Common Elder....................	Sambucus canadensis.
Condor Vine........................	Asclepias garciana.
Conedisk Sunflower..........	Rudbeckia laciniata.

LOCAL OR COMMON NAMES	BOTANICAL OR OFFICINAL NAMES.
Contrayerva	Dorstenia contrayerva.
Coral Root	Corallorhiza odontorhiza.
Cornea	Cornus Circinata.
Corn Snakeroot	*Eryngium aquaticum.*
" "	Liatris spicata.
Corpse Plant	Monotropa uniflora.
Cotton Root	Gossypium herbaceum.
Couch Grass	Triticum repens.
Cough Root	Trillium pendulum.
Counter Poison	Dorstenia contrayerva.
Cowbane	Cicuta virosa.
" Spotted	" maculata.
Cow Cabbage	Nymphæa odorata.
" Collard	Dracontium fœtidum.
" Lily	Nuphar advena.
" Parsnip	Heracleum lanatum.
Cowslips of Jerusalem	Pulmonaria officinalis.
Cranberry, High	*Viburnum opulus.*

LOCAL OR COMMON NAMES.	BOTANICAL OR OFFICINAL NAMES.
Cranberry, High	Arctostaphylos uva ursi.
" Mountain	" "
" Tree	Viburnum opulus.
" Wild	Arctostaphylos uva ursi.
" Upland	" " "
Crampback	*Viburnum opulus.*
"	Arctostaphylos uva ursi.
Cramp Bark	Viburnum opulus.
Cranesbill	Geranium maculatum.
" Spotted	" "
Crawley	Corallorhiza odontorhiza.
Creata	Agathotes chirayita.
Creeping Blackberry	Rubus trivialis.
Cress, Marsh	Nasturtium palustre.
" Water	" officinalis.
Crosswort	Eupatorium perfoliatum.
Crowcorn	Aletris farinosa.
Crowfoot	*Geranium maculatum.*

Local or Common Names.	Botanical or Officinal Names.
Crowfoot	Ranunculus bulbosus.
Crown Bark	Cinchona officinalis.
Cuckoo's Bread	Oxalis acetosella.
" Pint	Arum maculatum.
Cucumber, Indian	Medeola Virginica.
" Root	" "
" Tree	Liriodendron tulipifera.
Cudweed	Gnaphalium margaritaceum.
Culver's Physic	Leptandra Virginica.
" Root	" "
*Cundurango	Asclepias garciana.
Curcuma Root	Curcuma longa.
Cure-all	*Geum Virginianum.*
"	Melissa officinalis.
Curled Dock	Rumex crispus.
Cursed Thistle	Centaurea benedicta.
Cutting Almond	Parthenium integrifolium.
Cyprus Tree	Liriodendron tulipifera.

* No Officinal Name has yet been adopted by Botanists for Cundurango.

LOCAL OR COMMON NAMES.	BOTANICAL OR OFFICINAL NAMES.
Daffodil	Narcissus pseudo-narcissus.
Dandelion	Leontodon taraxacum.
Deadarsmart	Polygonum punctatum.
Deadly Nightshade	Atropa belladonna.
Deer Berry	Gaultheria procumbens.
" Weed	Cimicifuga racemosa.
Devil's Apple	Datura stramonium.
" Bit	*Chamælirium lutea.*
" "	Helonias dioica.
" "	Aletris farinosa.
" Bones	*Liatris spicata.*
" "	Dioscorea villosa.
" Shoestring	Tephrosia Virginiana.
Dewberry	Rubus trivialis.
Dielytra	Corydalis formosa.
Digitalis	Digitalis purpurea.
Dill Seed	Anethum graveolens.
" Weed	Anthemis cotula.
Dilly	*Anethum graveolens.*
"	Anthemis cotula.
Dittany, American	Cunila mariana.

LOCAL OR COMMON NAMES.	BOTANICAL OR OFFICINAL NAMES
Dittany, Mountain............	Cunila mariana.
Dock, Curled............	Rumex crispus.
" Prairie....................	Parthenium integrifolium.
" Sour............	Rumex crispus.
" Spatter	Nuphar advena.
" Water....................	Rumex aquaticus.
" Yellow....................	" crispus.
Dog Fennel......................	Anthemis cotula.
" Grass...................	Triticum repens.
" Tooth Violet..............	Erythronium Americanum.
" Tree............................	Cornus Florida.
Dogwood, Blueberried....	" sericea.
" Female............	" "
" Florida.............	" Florida.
" Flowering.......	" "
" Male Virginian....	" "
" Round-leaved	" circinata.

LOCAL OR COMMON NAMES.	BOTANICAL OR OFFICINAL NAMES
Dogwood, Silky-leaved...	Cornus sericea.
" Swamp.............	" "
Dogsbane................................	Apocynum androsæmifolium.
" Bitter..................	" "
Double Tansy.............	Tanacetum crispus.
Dove Foot...............	Geranium maculatum.
Dracontium.........................	Symplocarpus fœtidus.
Dragon's Claw....................	Corallorhiza odontorhiza.
" Head..................	Dracocephalum canariense.
" Root....................	Arum triphyllum.
" Turnip...............	" "
Drooping Starwort...........	Helonias dioica.
Dropsy Plant.......................	Melissa officinalis.
Dropwort.................................	Gillenia trifoliata.
Doctor Tinker's Weed.....	Triosteum perfoliatum.
Duck's Foot.........................	Podophyllum peltatum.
Dulcamara	Solanum dulcamara.
Dulce....................	Fucus.
Dutch Myrtle...........	Myrica gale.

LOCAL OR COMMON NAMES.	BOTANICAL OR OFFICINAL NAMES.
Dwale	Atropa belladonna.
Dwarf Elder	Aralia hispida.
Dyer's Saffron	Carthamus tinctoria.
" Alkanet	Anchusa "
Dysentery Root	Gnaphalium uglinosum.
East India Root	Galanga, major and minor.
Elder, Dwarf	Aralia hispida.
" Flowers	Sambucus canadensis.
" Wild	Aralia hispida.
Elecampane	Inula helenium.
Elm Bark, Slippery	Ulmus fulva.
" " Sweet	" "
Emetic Herb	Lobelia inflata.
" Weed	" "
Erythronium	Erythronium Americanum.
European Century	Eythræa centaurium.
Evan Root	Geum Virginianum.
Evening Primrose	Œnothera biennis.

Local or Common Names.	Botanical or Officinal Names.
Eve's Cup	Saracenia officinalis.
Eye Balm	Hydrastis canadensis.
" Bright	*Euphrasia officinalis.*
" "	Lobelia inflata.
Fairy's Glove	Digitalis purpurea.
False Alder	Prinos verticillatus.
" Aloe	*Aletris farinosa.*
" "	Agave Virginica.
" Bittersweet	Celastrus scandens.
" Coltsfoot	-Asarum canadense.
" Ipecac	Triosteum perfoliatum.
" Unicorn	Helonias dioica.
" White Cedar	Thuja occidentalis.
Featherfew	Pyrethrum parthenium.
Female Dogwood	Cornus sericea.
" Fern	Asplenium filix fœmina.
" Regulator	Senecio gracilis.
Fern Gale	Comptonia asplenifolia.

LOCAL OR COMMON NAMES.	BOTANICAL OR OFFICINAL NAMES.
Fever Bush..............	*Prinos verticellatus.*
" "	Benzoin odoriferum.
" " ....'...	Laurus benzoin.
Feverfew...............	*Chrysanthemum parthenium.*
"	Pyrethrum "
Fever Root.............	*Corallorhiza odontorhiza.*
" "	Triosteum perfoliatum.
" **Tree**.............	Pinckneya pubens.
" **Twig**.................	*Celastrus scandens.*
" "	Apocynum androsæmifolium.
" **Wood**...............	*Benzoin odoriferum.*
" "	Laurus benzoin.
" **Wort**...............	*Eupatorium perfoliatum.*
" "	Triosteum "
Fiddle-leaved Bindweed	Convolvulus panduratus.
Field Balsam................	Gnaphalium margaritaceum.
Field Larkspur...............	Delphinium consolida.
" **Weed**..................	Anthemis cotula.
Figwort..........................	Scrophularia nodosa.

LOCAL OR COMMON NAMES.	BOTANICAL OR OFFICINAL NAMES.
Fine-leaved Water Hemlock	Œnanthe phellandrium.
Fire Weed	*Senecio hieracifolia.*
" "	Erechthites "
Fishmouth	Chelone glabra.
Fit Plant	Monotropa uniflora.
Fit Root	" "
Five Fingers	*Potentilla canadensis.*
" "	Panax quinquefolia.
" Flowered Gentian	Gentiana quinqueflora.
Flag Lily	Iris versicolor.
Flag Root, Sweet	Acorus calamus.
Flake	Anthemis cotula.
Fleabane, Canada	Erigeron canadense.
" Philadelphia	" Philadelphicum.
" Various-colored	" heterophyllum.
Fleur-de-lis	Iris versicolor.
Fluellin	Veronica officinalis.

LOCAL OR COMMON NAMES.	BOTANICAL OR OFFICINAL NAMES.
Florida Bark	Pinckneya pubens.
" Dogwood	Cornus Florida.
Flowering Dogwood	" "
Flower Velure	Tussilago farfara.
Flux Root	*Asclepias tuberosa.*
" "	Gentiana catesbæi.
Fly Trap	*Saracenia purpurea.*
" "	Apocynum androsæmifolium.
Fœtid Nightshade	Hyoscyamus niger.
Foreign Indian Hemp	Cannabis indica.
" Saffron	Crocus sativus.
Foxglove	Digitalis purpurea.
Frasera	Frasera walteri.
Frog Lily	Nuphar advena.
Frost Weed	*Helianthemum canadense.*
" "	Aster puniceus.
" Wort	*Helianthemum canadense.*
" "	Aster puniceus.
Fumitory	Fumaria officinalis.

LOCAL OR COMMON NAMES.	BOTANICAL OR OFFICINAL NAMES.
Galangal	Galanga, major and minor.
Gale, Sweet	Myrica gale.
Gall of the Earth	Gentiana quinquefolia.
Ganga	Cannabis indica.
Garantogen	Panax quinquefolia.
Garden Carrot	Daucus carota.
" Dill	Anethum graveolens.
" Nightshade	Solanum dulcamara.
" Patience	Rumex crispus.
" Rue	Ruta graveolens.
" Thyme	Thymus vulgaris.
Garget	Phytolacca decandra.
Gaultheria	Gaultheria procumbens.
Gay Feather	Liatris spicata.
Gensang	Panax quinquefolia.
Gentian	Gentiana lutea.
" Blue	" catesbæi.
" Great Yellow	" lutea.

LOCAL OR COMMON NAMES.	BOTANICAL OR OFFICINAL NAMES.
Gentian, White	Triosteum perfoliatum.
Georgia Bark	Pinkneya pubens.
German Chamomile	Matricaria chamomilla.
Gill go over the Ground	*Glechoma hederacea.*
" " " "	Nepeta glechoma.
Gill Run	" "
Ginger	Zingiber officinalis.
" Canada	Asarum canadense.
" Wild	" "
Ginseng	Panax quinquefolia.
Gipseywort	Lycopus Europeus.
Glechoma	Glechoma hederacea.
Goat's Rue	Tephrosia Virginiana.
Golden Fern	Aspidium filix mas.
" Ragwort	Senecio aurens.
" Rod	Solidago odora.
" Seal	*Hydrastis canadensis.*
" "	Frasera verticillata.

LOCAL OR COMMON NAMES.	BOTANICAL OR OFFICINAL NAMES.
Golden Senecio....................	Senecio aurens.
Gold Thread........................	Coptis trifolia.
Goose Grass........................	Galium aparine.
Gravel Laurel.............	Epigæa repens.
" Plant	*Epigæa repens.*
" "	Galium aparine.
" Weed......................	Epigæa repens.
" Root......................	Eupatorium purpureum.
Great American Century	Centaurea Americana.
" Flowering Dogwood	Cornus Florida.
" Water Dock.............	Rumex aquaticus.
" Wild Valerian........	Cypripedium pubescens.
" Yellow Dock...........	Rumex aquaticus.
" Yellow Gentian......	*Gentiana lutea.*
" " "	Frasera walteri.
Greek Valerian....................	Polemonium reptans.
Green Archangel................	Lycopus Europæus.

LOCAL OR COMMON NAMES.	BOTANICAL OR OFFICINAL NAMES.
Green Ozier	Cornus circinata.
Ground Berry	Gaultheria procumbens.
" Flower	Polygala rubella.
" Holly	*Chimaphilla umbellata.*
" "	Pyrola "
" Ivy	*Glechoma hederacea.*
" "	Gaultheria procumbens.
" Laurel	Epigæa repens.
" Leaf	Chimaphilla umbellata.
" Lily	Trillium pendulum.
" Maple	Heuchera Americana.
" Moss	Polytrichum juniperinum.
" Pine	Ajuga chamæpithys.
" Pink	Silene Virginica.
" Raspberry	Hydrastis canadensis.
" Sell	Senecio vulgaris. .
Grouse Berry	Gaultheria procumbens.
Guelder Rose	Viburnum opulus.
Guinea Pods	Capsicum annuum.

LOCAL OR COMMON NAMES.	BOTANICAL OR OFFICINAL NAMES
Gum Plant..............	*Pulmonaria officinalis.*
" "	Symphytum "
Gypsey Weed..........	Lycopus Virginicus.
Hair Cap Moss.........	Polytrichum juniperinum.
Hairy American Bramble	Rubus villosus.
" Sweet Cicily......	Osmorrhiza brevistylis.
Hardhack................	*Spiræa tomentosa.*
"	Collinsonia canadensis.
Heal-all................	*Scrophularia nodosa.*
"	Collinsonia canadensis.
"	Prunella Pennsylvanica.
Healing Herb...........	*Symphytum officinale.*
" "	Pulmonaria officinalis.
Heart Root.............	Asarum canadense.
" Snakeroot........	" "
Hedge Mustard..............	Sisymbrium officinalis.
Hellebore, American.......	Veratrum viride.
" Black..............	Helleborus niger.
" Fœtid	Dracontium fœtidum.
" Swamp............	Helleborus Americana.
" White..............	Veratrum album.

LOCAL OR COMMON NAMES.	BOTANICAL OR OFFICINAL NAMES.
Hemlock, Fine-leaved......	Œnanthe phellandrium.
" **Poison**	Conium maculatum.
" **Spruce**	Pinus canadensis.
" **Water Drop**.....	*Conium maculatum.*
" "	Œnanthe crocata.
Hemp, Foreign..................	Cannabis indica.
" **Indian**.....................	Apocynum cannabinum.
" **White**	Asclepias incarnata.
Henbane, Black..................	Hyoscyamus niger.
Herb de Paigne..................	Chimaphila umbellata.
Herb Robert....................	Geranium robertianum.
Hercules' Club..................	Aralia spinosa.
Hibiscus............................	Althæa officinalis.
High Angelica....................	Angelica atropurpurea.
" **Blackberry**	Rubus villosus.
" **Cranberry**	Viburnum opulus.
" **Mallows**	Malva sylvestris.

LOCAL OR COMMON NAMES.	BOTANICAL OR OFFICINAL NAMES.
Hive Vine	Mitchella repens.
Hoarhound	Marrubium vulgare.
" White	" "
Hoary Pea	Tephrosia Virginiana.
Hog Potato	Convolvulus panduratus.
" Weed	Ambrosia artemisæfolia.
Hollyhock	Althæa rosea.
Holly Rose	Cistus canadensis.
Holy Thistle	Centaurea benedicta.
Honey Bloom	Apocynum androsæmifolium.
Hooded Willow Herb	Scutellaria lateriflora.
Hoodwort	" "
Hops	Humulus lupulus.
Hop Tree	Ptelea trifoliata.
Horse Balsam	Collinsonia canadensis.
" Chestnut, English	Æsculus hippocastaneum.
" " Ohio	" glabra.
" Fly Weed	Baptisia tinctoria.
" Gentian	Triosteum perfoliatum.
" Mint	Monarda punctata.

LOCAL OR COMMON NAMES.	BOTANICAL OR OFFICINAL NAMES.
Horse Radish...............	Cochlearia armoracia.
" Weed............ .	*Collinsonia canadensis.*
" "	Erigeron canadense.
Hound's Tongue........	Cynoglossum officinale.
Humulus..................	Humulus lupulus.
Huntsman's Cup................	Saracenia purpurea.
Hydrangea........	Hydrangea arborescens.
Hydrolapathrum	Rumex aquaticus.
Hyssop.........................	Hyssopus officinalis.
Ibiscus..................	Althæa "
Iceland Moss............	Cetraria Islandica.
Ice Plant..	Monotropa uniflora.
" Vine................................	Cissampelos pareira.
Impatiens........	Impatiens balsamina.
Indian Apple........................	Podophyllum peltatum.
" Arrow-wood.........•	Euonymus atropurpureus.
" Balm.....................	Trillium pendulum.
" Cucumber	Medeola Virginica.
" Dream.............	Pteris atropurpurea.
" Elm..............	Ulmus fulva.
" Ginger........	Asarum canadensis.

LOCAL OR COMMON NAMES.	BOTANICAL OR OFFICINAL NAMES.
Indian Hemp	Apocynum cannabinum.
" " **Foreign**	Cannabis indica.
" " **White**	Asclepias incarnata.
" **Hippo**	Gillenia trifoliata.
" **Lettuce**	*Frasera walteri.*
" "	Cocculus palmatus.
" **Paint**	*Sanguinaria canadensis.*
" "	Hydrastis canadensis.
" **Physic**	Gillenia trifoliata.
" **Pink**	Lobelia cardinalis.
" " **Root**	Spigelia marilandica.
" **Pipe**	Monotropa uniflora.
" **Poke**	Helleborus Americana.
" **Posey**	Gnaphalium polycephalum.
" **Sage**	Eupatorium perfoliatum.
" **Senna**	Cassia marilandica.
" **Shoe**	Cypripedium pubescens.
" **Tobacco**	Lobelia inflata.

LOCAL OR COMMON NAMES.	BOTANICAL OR OFFICINAL NAMES
Indian Turnip	Arum triphyllum.
Indigo Broom	Baptisia tinctoria.
" Weed	" "
" Wild	" "
Indigofera	" "
Ink Root	Statice Caroliniana.
Ipecac	*Cephælis ipecacuanha.*
'	Podophyllum peltatum.
" American	Euphorbia ipecacuanha.
" Milk	Apocynum androsæmifolium.
" Spurge	Euphorbia ipecacuanha.
" Wild	Triosteum perfoliatum.
Ipecacuanha	*Cephælis ipecacuanha.*
"	Podophyllum peltatum.
Irish Cabbage	Dracontium fœtidum.
Itch Weed	*Veratrum viride.*
"	Helleborus Americana.

LOCAL OR COMMON NAMES.	BOTANICAL OR OFFICINAL NAMES
Itch Weed............................	Verbascum thapsus.
"	Symplocarpus fœtidus.
Ivy....................................	Kalmia latifolia.
" Ground...........................	Glechoma hederacea.
Jack in the Pulpit.............	Arum triphyllum.
Jacob's Ladder.....................	*Smilax peduncularis.*
" "	Polemonium reptans.
" "	Pulmonaria officinalis.
Jalap................................	Ipomæa vulgaris.
" Cancer Root...............	Phytolacca decandra.
Jamestown Weed..............	Datura stramonium.
Jersey Tea..............	Ceanothus Americana.
Jerusalem Oak.........	Chenopodium anthelminticum.
Jessamine, Carolina.....	Gelseminum sempervirens.
" Yellow......	" "
Jesuit's Bark...........	Cinchona officinalis.
" Tea........	Chenopodium anthelminticum.
Jewel Weed............	Impatiens fulva and pallida.
Jewsharp................	Trillium pendulum.

LOCAL OR COMMON NAMES.	BOTANICAL OR OFFICINAL NAMES.
Jimpson Weed	Datura stramonium.
Joe Pye	Eupatorium purpureum.
Johnswort	Hypericum perforatum.
Juniper	Juniperus communis.
Kassamah	Galanga major and minor.
Kidney Liver Leaf	Hepatica Americana.
Kidney-leaved Asara-bacca	Asarum canadense.
King's Clover	*Melilotus officinalis.*
" "	Melissa officinalis.
" Cure	*Chimaphila umbellata.*
" "	Pyrola umbellata.
" "	" maculata.
Knobs Grass	Collinsonia canadensis.
Knot "	Triticum repens.
Knot Root	Collinsonia canadensis.
" Weed	Polygonum punctatum.
Knotty-rooted Figwort	Scrophularia nodosa.
Kusawda	Convolvulus panduratus.
Lady's Slipper	*Cypripedium pubescens.*

LOCAL OR COMMON NAMES.	BOTANICAL OR OFFICINAL NAMES.
Lady's Slipper	Impatiens balsamina.
" "	Variolaria faginica.
" "	Valeriana officinalis.
Lad's Love	Artemisia abrotanum.
Lamb-kill	Kalmia latifolia.
" Quarter	Trillium pendulum.
Large Flowering Cornel	Cornus Florida.
" " Spurge	Euphorbia corollata.
Large Yellow Lady's Slipper	Cypripedium pubescens.
Larkspur	*Delphinium consolida.*
"	" staphisagria.
" Branching	" "
" Field	" "
Laurel, Broad-leaved	Kalmia latifolia.
" Mountain	" "
" Sheep's	" angustifolia.
Laurel, Sheep's, Narrow-leaved	" "
Laurel, Swamp	" glauca.

LOCAL OR COMMON NAMES.	BOTANICAL OR OFFICINAL NAMES.
Laurus	Laurus nobilis.
Lavender	Lavendula vera.
Lavose	Ligusticum levisticum.
Lemon Balm	Melissa officinalis.
" Walnut	Juglans cinerea.
" Wild	Podophyllum peltatum.
Len-nik-bi	Liriodendron tulipifera.
Leontice	Caulophyllum thalictroides.
Leopard-bane	Arnica montana.
Lettuce, Garden	Lactuca sativa.
" Wild	" elongata.
Licorice	Glycyrrhiza glabra.
Life Everlasting	Gnaphalium polycephalum.
Life of Man	*Aralia racemosa.*
" "	Nard.
" Root	Senecio aureus.
Lily of the Valley	Convallaria majalis.

LOCAL OR COMMON NAMES	BOTANICAL OR OFFICINAL NAMES.
Linaria	Antirrhinum linaria.
Lion's Foot	Nabalus alba.
Liquorice	Glycyrrhiza glabra.
Live for Ever	Gnaphalium polycephalum
Liver Leaf	Hepatica Americana.
" Lily	Iris versicolor.
Liverwort	*Hepatica Americana.*
"	" triloba.
Lobelia	Lobelia inflata.
" Red	" cardinalis.
" Blue	" syphilitica.
Locust Plant	*Cassia marilandica.*
" "	" acutifolia.
London Pride	Saponaria officinalis
Long-leaved Water Parsnip	Cicuta virosa.
Loose-strife	Lythrum salicaria.
Lousewort	Pedicularis canadensis.
Lovage	Ligusticum levisticum.
Lovely Thistle	Centaurea benedicta.

LOCAL OR COMMON NAMES.	BOTANICAL OR OFFICINAL NAMES.
Love Lies Bleeding.....	Amaranthus melancholicus.
Low Chamomile...............	Anthemis nobilis.
" Cud Weed..................	Gnaphalium uliginosum.
" Mallows..................	Malva rotundifolia.
Lungwort......................	*Pulmonaria officinalis.*
"	Variolaria faginica.
Lupulin................	Humulus lupulus.
Lyre-leaved Sage.......	Salvia lyrata.
Lyre Tree of America.......	Liriodendron tulipifera.
Mad Dog Scull-cap............	Scutellaria lateriflora.
Mad Weed............................	" "
Magwort..............................	Artemisia vulgaris.
Maiden Hair........................	Adiantum pedatum.
Male Fern...........................	Aspidium filix-mas.
" Shield Fern...............	" "
" Virginia Dogwood ..	Cornus Florida.
Mallows, Common.............	Malva sylvestris.
" Low......................	" rotundifolia.

LOCAL OR COMMON NAMES.	BOTANICAL OR OFFICINAL NAMES
Mallows, Marsh...............	Althæa officinalis.
Mandrake......................	Podophyllum peltatum.
Man in the Earth............	Convolvulus panduratus.
" " Ground...........	" "
" of the Earth.............	" "
" Root.......................	" "
Maple-leaf Alum Root......	Heuchera Americana.
" · Lungwort...............	Pulmonaria officinalis.
Mare's Tail...................	Erigeron canadense.
Marietta Colombia...........	Frasera walteri.
Marigold.....................	Calendula officinalis.
Marjoram, Sweet.............	Origanum majorana.
" Wild.................	" vulgare.
Marsh Mallows...............	Althaea officinalis.
" Rosemary.............	Statice Caroliniana.
" Trifoil	Menyanthes trifoliata.
" Turnip	Arum triphyllum.

LOCAL OR COMMON NAMES.	BOTANICAL OR OFFICINAL NAMES.
Marsh Water Cress...........	Nasturtium palustre.
Marubian...............................	Marrubium vulgare.
Maryland Cassia.................	Cassia marilandica.
" Cunila.................	Cunila mariana.
Masterwort...........................	*Angelica atropurpurea.*
"	Imperatoria ostruthium.
Masterwort, leaves, root and seed.	Heracleum lanatum.
Mata Perro..........................	Asclepias garciana.
" "	" mata perro.
Mathen	Anthemis cotula.
Matico....................................	*Piper angustifolium.*
"	Artanthe elongata.
May Apple............................	Podophyllum peltatum.
May Flower...........................	*Anthemis cotula.*
" "	Epigæa repens.
" Weed...........................	Anthemis cotula.
" " Flower...............	" ".

LOCAL OR COMMON NAMES.	BOTANICAL OR OFFICINAL NAMES.
Meadow Anemone......	Anemone pratense.
" **Bloom**.........	*Ranunculus acris.*
" "	Cimicifuga racemosa.
" **Cabbage**.......	Dracontium fœtidum.
" **Crowfoot**......	*Ranunculus acris.*
" "	Symplocarpus fœtidus.
" **Fern**...........	*Comptonia asplenifolia.*
" "	Frasera walteri.
" " ..·........	Myrica gale.
" **Pride**..........	Frasera walteri.
" **Root**...........	Statice Caroliniana.
" **Saffron**........	Colchicum autumnale.
" **Sage**..........	Salvia lyrata.
" **Scabish**........	Aster puniceus.
" **Sweet**.........	Gillenia trifoliata.
" "	*Spirea tomentosa.*
Mealy Starwort...............	Aletris farinosa.
Mechamech.........................	Convolvulus panduratus.
Mecoama.............................	" "
Melissa.............................	Melissa officinalis.
Mellilot.............................	Melilotus officinalis.
Mezereun...........................	Daphne mezereum.

LOCAL OR COMMON NAMES.	BOTANICAL OR OFFICINAL NAMES.
Milfoil	Achillea millefolium.
Milksweet	Galium aparine.
Milk Weed	*Apocynum androsæmifolium.*
" "	Euphorbia corollata.
" " Root	Asclepias incarnata.
" " Wandering	Asclepias syriaca.
Milkwort	Polygala senega.
Mint-leaved Cunila	Cunila mariana.
Moccasin Root	Cypripedium pubescens.
Moldavian Balm	Dracocephalum canariense.
Monkey Flower	Mimulus moscatus.
Monkshood	*Aconitum napellus.*
"	Leontodon taraxacum.
Mortification Root	Althaea officinalis.
Mother of Thyme	Thymus vulgaris.
Motherwort	Leonurus cardiaca.
Mountain Balm	Monarda didyma.
" Box	Arctostaphylos uva ursi.

LOCAL OR COMMON NAMES.	BOTANICAL OR OFFICINAL NAMES.
Mountain Cranberry......	Arctostaphylos uva ursi.
" Dittany...........	Cunila mariana.
" Flax................	Polygala senega.
" Laurel.............	Kalmia latifolia.
" Mint...............	*Origanum vulgare.*
" "	Monarda didyma.
" Pink...............	Epigæa repens.
" Tea................	Gaultheria procumbens.
" Tobacco...........	Arnica montana.
" Zarusum.........	Simaruba excelsa.
Mouse Ear...................	Gnaphalium uliginosum.
Mouth Root.................	Coptis trifolia.
Mugwort....................	Artemisia vulgaris.
Mullein.....................	Verbascum thapsus.
Musk Plant.............	Mimulus moscatus.
Musquash Root.........	Cicuta maculata.
Mustard, Hedge........	Sisymbrium officinalis.
Myrtle Flag............	Acorus calamus.

LOCAL OR COMMON NAMES.	BOTANICAL OR OFFICINAL NAMES.
Myrtle Leaves	Myrica cerifera.
Nanny Bush Bark	Viburnum opulus.
Narcissus	Narcissus pseudo-narcissus.
Nardus	Nard.
Narrow Dock	Rumex crispus.
Narrow-leaved Piper	Piper angustifolium.
Narrow-leaved Sheep's Poison	Kalmia "
Narrow-leaved Sheep's Laurel	" "
Narrow-rayed Robin's Plantain	Erigeron Philadelphicum.
Nasturtion	Tropæolum majus.
Necklace Weed	Actea alba.
Nectarine	Amygdalus persica
Nepeta	Nepeta cataria.
Nephritic Plant	Parthenium integrifolium.
Nerve Root	Cypripedium pubescens.
" " Water	" "

LOCAL OR COMMON NAMES.	BOTANICAL OR OFFICINAL NAMES.
Nervine.................	Cypripedium pubescens.
Net-leaf Plantain........	Goodyeara pubescens.
Nettle-leaved Vervain..	Verbena urticifolia.
Nettle Root	Urtica dioca.
New England Box Wood	*Cornus Florida.*
" " "	Rhus glabra.
New Jersey Tea.........	Ceanothus Americanus.
Nightshade....................	Atropa belladonna.
Ninsin...................	Panax quinquefolia.
Noah's Ark....................	*Cypripedium pubescens.*
" "	Actea alba.
Noble Pine....................	Chimaphila umbellata.
Nymph....................	Nymphæa odorata.
Ohio Buckeye...........	Æsculus glabra.
" Curcuma...........	*Hydrastis canadensis.*
" "	Frasera verticillata.
Oil Nut.................	Juglans cinerea.
Old Maid's Pink....	Saponaria officinalis.
" Man................	Artemisia abrotanum.
" Wife's Shirt........	Liriodendron tulipifera.
Olive Spurge...........	Daphne mezereum.
One Berry...............	Mitchella repens.

LOCAL OR COMMON NAMES.	BOTANICAL OR OFFICINAL NAMES.
Opium Poppy......................	Papaver somniferum.
Orange Apocynum.............	Asclepias tuberosa.
" Peel........................	Aurantii cortex.
" Root........................	Hydrastis canadensis.
" Swallow Root......	Asclepias tuberosa.
Orris Root............................	Iris florentina.
Oswego Tea.........................	Monarda didyma.
Ova-ova..................	Monotropa uniflora.
Ox Balm................	Collinsonia canadensis.
Pæony.	Pæonia officinalis.
Pappoose Root...........	*Caulophyllum tha'ictroides.*
" "	Leontice "
Pareira Brava	Cissampelos pareira.
Parilla, Yellow.........	Menispermum canadense.
Pariswort...............	Trillium pendulum.
Parsley Leaves and Seed	Apium petroselinum.
" Root	" "
" Yellow Root ...	Zanthoxylum fraxineum.

LOCAL OR COMMON NAMES.	BOTANICAL OR OFFICINAL NAMES.
Parsnip, Water	Sium nodiflorum.
Partridge Berry	*Gaultheria procumbens.*
" "	Momordica balsamina.
" Vine	Mitchella repens.
Pasque Flower	Anemone pratense.
Paul's Betony	Lycopus Virginicus.
Peach Leaves	Amygdalus persica.
Pearly Everlasting	Gnaphalium margaritaceum.
Pecatacalleloe	Phytolacca decandra.
Pelitory	*Anacyclus pyrethrum.*
"	Rhus glabra.
" of Spain	Anacyclus pyrethrum.
Pennsylvania Sumach	Xanthoxylum fraxineum.
Pennyroyal	Hedeoma pulegioides.
Peony	Pæonia officinalis.
Peppermint	Mentha piperita.
Pepper Turnip	Arum triphyllum.

LOCAL OR COMMON NAMES.	BOTANICAL OR OFFICINAL NAMES.
Pepper, Water	Polygonum punctatum.
Persoon	Pyrethrum parthenium.
Peruvian Bark	Cinchona officinalis.
Petty Morel	*Aralia racemosa.*
" "	Nard.
Philadelphia Fleabane	Erigeron Philadelphicum.
Physic, Culvers	Leptandria Virginica.
" Indian	*Gillenia trifoliata.*
" "	Phytolacca decandra.
Pigeon Berry	" "
Pilewort	*Amaranthus hypochondriachus.*
"	Ranunculus acris.
Pimpernel, Scarlet	Anagallis arvensis.
Pinckneya	Pinckneya pubens.
Pine, Ground	Ajuga chamæpithys.
" Noble	Chimaphila umbellata.
" Prince's	" "
" Sap	Monotropa uniflora.
" Tulip	Cypripedium pubescens.

LOCAL OR COMMON NAMES.	BOTANICAL OR OFFICINAL NAMES.
Pink, Wild	Silene Virginica.
Pink Root, Carolina	Spigelia marilandica.
Pinkneya	Pinckneya pubens.
Pipe Plant	Monotropa uniflora.
Pipperidge	Berberis vulgaris.
Pipsissewa	Chimaphila umbellata.
Pitcher Plant	Saracenia officinalis.
Plantain	Plantago major.
" Greater	" "
" Water	Alisma plantago.
" Bitter	Gentiana quinquefolia.
Pleurisy Root	Asclepias tuberosa.
Poison Ash	Rhus toxicodendron.
" Flag	Iris versicolor.
" Hemlock	Conium maculatum.
" Ivy	Rhus toxicodendron.
" Oak	" "
" Parsley	Conium maculatum.

LOCAL OR COMMON NAMES.	BOTANICAL OR OFFICINAL NAMES.
Poison Tobacco	Hyoscyamus niger.
" Vine	Rhus toxicodendron.
Poke	*Phytolacca decandra.*
"	Dracontium fœtidum.
" Indian	Helleborus Americanus.
Polecat Weed	Dracontium fœtidum.
" Collard	" "
Pollypody, Common	Polypodium vulgare.
Polygala	Polygala rubella.
Pool Root	*Eupatorium aromaticum.*
" "	Sanicula marilandica.
Poor Man's Weather-glass	Anagallis arvensis.
Poor Robin	Galium aparine.
Poplar, American	Populus tremuloides.
Poppy, Opium	Papaver somniferum.
Prairie Dock	Parthenium integrifolium.

LOCAL OR COMMON NAMES.	BOTANICAL OR OFFICINAL NAMES.
Prassium	Marrubium vulgare.
Prickly Ash	*Zanthoxylum fraxineum.*
" "	Aralia nudicaulis.
" Elder	" spinosa.
Pride of China	Melia azedarach.
" India	" "
" Weed	Erigeron canadense.
Prim	Ligustrum vulgare.
Primrose Tree	Œnothera biennis.
Prince's Feather	Amaranthus hypochondriachus
" Pine	Chimaphila umbellata.
Privet	Ligustrum vulgare.
Puccoon, Red	Sanguinaria canadensis.
" Yellow	Hydrastis canadensis.
Puke Root	Lobelia inflata.
Pulsatilla	Anemone pulsatilla.
Pumpkin Seed	Cucurbita pepo.
Purging Berries	Rhamnus catharticus.
Purple Angelica	Angelica atropurpurea.

LOCAL OR COMMON NAMES.	BOTANICAL OR OFFICINAL NAMES.
Purple Avens...............	Geum rivale.
" Boneset..................	Eupatorium purpureum.
" Foxglove	Digitalis purpurea.
" Willow Herb........	Lythrum salicaria.
Purvain..............................	Verbena hastata.
Pussy Willow...........	Salix nigra.
Putty Root..............	Corallorhiza hyemalis.
Pyramid Flower........	*Frasera walteri.*
" "	Cocculus palmatus.
Pyrethrum Root........	Anacyclus pyrethrum.
Pyrola..................	*Chimaphilla umbellata.*
"	Pyrola "
Quaking Aspen..........	Populus tremuloides.
Quassia.................	Simaruba excelsa.
Queen of the Meadow..	*Eupatorium purpureum.*
" " " ..	Spiræa ulmaria.
Queen's Delight.........	Stillingia sylvatica.
" Root...........	" "
Quick Grass....	Triticum repens.
Quicken.................	" "
Quince Seed............	Cydonia vulgaris.

LOCAL OR COMMON NAMES.	BOTANICAL OR OFFICINAL NAMES.
Quitch Grass	Triticum repens.
Quiver Leaf	Populus tremuloides.
Racine a Becquet	Geranium maculatum.
Raccoon Berry	Podophyllum peltatum.
Radish, Water	Nasturtium amphibium.
Rag Weed	Ambrosia artemisæfolia.
Raspberry, Ground	Hydrastis canadensis.
" Leaves	Rubus strigosus.
Rattlebush	Baptisia tinctoria.
Rattle Root	*Macrotys racemosa.*
" "	Cimicifuga racemosa.
" "	Cicuta virosa.
Rattlesnake Root	*Goodyara pubescens.*
" "	Nabulus alba.
" "	Cimicifuga racemosa.
" "	Polygala senega.
" "	Liatris spicata.
" " Violet..	*Erythronium Americanum.*

Local or Common Names.	Botanical or Officinal Names.
Rattlesnake Violet.............	Liatris squarrosa.
Rattlesnake's Master........	*Eryngium aquaticum.*
" "	Agave Virginica.
" "	Liatris squarrosa.
" "	Nabulas alba.
Red Alder...................	Alnus rubra.
" Archangel..................	Lycopus Virginicus.
" Balm.....................	Monarda didyma.
" Bark.....................	Cinchona rubra.
" Berry....................	*Gaultheria procumbens.*
" "	Arctostaphylos uva ursi.
" "	Panax quinquefolia.
" Blood Root..............	Sanguinaria canadensis.
" Cardinal Flower.......	Lobelia cardinalis.
" " Plant..........	" "
" Cedar....................	Juniperus Virginiana.
" Century.................	Sabbatia angularis.
" Clover...................	Trifolium pratense.
" Cockscomb..............	Amaranthus hypochondriachus

LOCAL OR COMMON NAMES.	BOTANICAL OR OFFICINAL NAMES
Red Cohosh	Actea rubra.
" Elm	Ulmus fulva.
" Lobelia	Lobelia cardinalis.
" Oak Bark	Quercus rubra.
" Osier	Cornus sericea.
" Pepper	Capsicum annuum.
" Puccoon	Sanguinaria canadensis.
" Rod	Cornus sericea.
" Root	Sanguinaria canadensis.
" " Bark	Ceanothes Americana.
" Stalked Aster	Aster puniceus.
" Tag Alder	Alnus rubra.
" Willow	Cornus sericea.
Rhatany	Krameria triandra.
Rheumatism Weed	*Chimaphila maculata.*
" "	Aster puniceus.
Rib Grass	Plantago major.
Rich Weed	*Collinsonia canadensis.*
" "	Cimicifuga racemosa.

LOCAL OR COMMON NAMES.	BOTANICAL OR OFFICINAL NAMES.
Robin Run the Hedge......	*Galium aparine.*
" " "	Nepeta glechoma.
Robin's Rye................	Polytrichum juniperinum.
Rock Fern..................	Adiantum pedatum.
Rockbrake...	Pteris atropurpurea.
Rock Parsley...............	Petroselinum sativum.
" Polypody............	Polypodium vulgare.
" Rose...............	*Cistus canadensis.*
" "	Helianthemum canadense.
Roman Chamomile............	Anthemis nobilis.
" Wormwood...........	Ambrosia artemisæfolia.
Rose-colored Silk Weed	Asclepias incarnata.
Rosemarinus...............	Rosmarinus officinalis.
Rosemary.................	" "
" Marsh...............	Statice Caroliniana.
Rose Pink................	Centaurea Americana.
" "	*Sabbatia angularis.*

LOCAL OR COMMON NAMES.	BOTANICAL OR OFFICINAL NAMES.
Rose Willow	Cornus sericea.
Round-leaved Consumption Plant	Pyrola rotundifolia.
Round-leaved Dogwood	Cornus circinata.
" " Pyrola	Pyrola umbellata.
Royal Ocymum	Ocymum basilicum.
Rue	Ruta graveolens.
Rum Cherry	Prunus Virginiana.
Sabbatia	Sabbatia angularis.
Safflower	Carthamus tinctorius.
Saffron, American	" "
" Foreign	Crocus sativus.
" Spanish	" "
Sage	Salvia officinalis.
Sage of Jerusalem	Pulmonaria officinalis.
Saint Johnswort	Hypericum perforatum.
Salt Rheum Weed	Chelone glabra.
Sambucus	Sambucus canadensis.

LOCAL OR COMMON NAMES.	BOTANICAL OR OFFICINAL NAMES.
Sampscus Majoram............	Origanum majorana.
Sampson Snake Root........	Gentiana saponaria.
Sangrel............................	Aristolochia serpentaria.
Sanicle...........................	Cinchona officinalis.
" Black......................	Sanicula marilandica.
Sarsaparilla, Honduras...	Smilax officinalis.
" American...	Aralia nudicaulis.
Sassafras........................	*Laurus sassafras.*
"	Sassafras officinalis.
" Bark...................	Sassafras radicis cortex.
Savine...........................	Juniperus sabina.
Savoyan.........................	Galium aparine.
Saxafrax........................	Laurus sassafras.
Saxifrage.......................	Saxifraga pimpinella.
Scabish.........................	Œnothera biennis.
"	*Erigeron Philadelphicum*
Scabius........................	" "

LOCAL OR COMMON NAMES.	BOTANICAL OR OFFICINAL NAMES.
Scabius, Sweet	Erigeron Philadelphicum.
Scabwort	Inula helenium.
Scarlet Berry	Solanum dulcamara.
Scarlet Pimpernel	Anagallis arvensis.
Sclarea	Salvia sclarea.
Scoke Root	Phytolacca decandra.
Scrofula Plant	Scrophularia lateriflora.
" Weed	Helianthemum canadense.
Scullcap	Scutellaria lateriflora.
Scurvy Grass	Cochlearia officinalis.
Sea Ash	Aralia nudicaulis.
" Lavender	Statice Caroliniana.
Seaside Balsam	Croton eleuteria.
Seathrift	Statice Caroliniana.
Self Heal	Prunella Pennsylvanica.
Senega	Polygala senega.
Seneka Snake Root	" "

LOCAL OR COMMON NAMES.	BOTANICAL OR OFFICINAL NAMES.
Senna, Alexandria	Cassia acutifolia.
" American	" marilandica.
" Wild	" "
Serpentaria	Aristolochia serpentaria.
Setterswort	Helleborus fœtidus.
Seven Barks	Hydrangea arborescens.
Sheep Berry	Viburnum lentago.
" Laurel	Kalmia latifolia.
" Poison	" "
" Sorrel	Oxalis acetosella.
Shepherd's Purse	Thlaspi bursa pastoris.
Shin Leaf	Pyrola rotundifolia.
Short Styled Sweet Cicily	Osmorrhiza brevistylis.
Shotbush	Aralia nudicaulis.
Shrub Yellow Root	Zanthoriza apiifolia.
Shrubby Bittersweet	Celastrus scandens.
" Sweet Fern	Comptonia asplenifolia.

LOCAL OR COMMON NAMES	BOTANICAL OR OFFICINAL NAMES.
Shrubby Trefoil................	Ptelea trifoliata.
Side-saddle Plant...............	Sarracenia purpurea.
Silk Weed Root........	Asclepias syriaca.
Silky Cornel.............	Cornus sericea.
Silky-leaved Dogwood...	Cornus sericea.
Silver Leaf........................	Stillingia sylvatica.
Simaruba............................	Simaruba excelsa.
Simplers Joy.......................	Verbena hastata.
Skavish...............................	Erigeron purpureum.
Skoka..................................	*Dracontium fœtidum.*
"	Phytolacca decandra.
Skoke Root.........................	" "
Skunk Cabbage...................	*Dracontium fœtidum.*
" "	Symplocarpus fœtidus.
" Weed......................	" "
Slender-spiked Liatris.....	Liatris spicata.
Slippery Elm.............	Ulmus fulva.
Sloe....................	Viburnum prunifolium.
Small Burnet Saxifrage..	Saxifraga pimpinella.
" Pox Plant........	Saracenia purpurea.

LOCAL OR COMMON NAMES.	BOTANICAL OR OFFICINAL NAMES.
Smart Weed...........................	Polygonum hydropiper.
Smellage	Ligusticum levisticum.
Smooth Alder........................	Alnus rubra.
" Sumach....................	Rhus glabra.
Snakebite...............	*Lactuca elongata.*
"	Trillium pendulum.
Snakehead...............	Chelone glabra.
Snake Lilly...............	Iris versicolor.
" Root, Black.......	Cimicifuga racemosa.
" " "	Macrotys
" " Birthwort...	Aristolochia serpentaria.
" " Button..........	*Eryngium aquaticum.*
" " "	Liatris spicata.
" " Canada.........	Asarum canadensis.
" " Corn.............	*Eryngium aquaticum.*
" " "	Liatris spicata.
" " Heart.......	Asarum canadense.
" " Senega......	Polygala senega.
" " Seneka......	" "
" " Virginia....	Aristolochia serpentaria.

LOCAL OR COMMON NAMES.	BOTANICAL OR OFFICINAL NAMES.
Snake Root, White......	*Eupatorium aromaticum.*
" " "	" urticifolium.
" " "	Aristolochia serpentaria.
" " Wild........	Asarum canadens...
" Weed.............	*Aristolochia serpentaria.*
" "	Cicuta maculata.
" "	Arctostaphylos uva ursi
" "	Polygala senega.
" "	Lactuca elongata.
Snapping Hazelnut......	Hamamelis Virginica.
Snap Dragon............	Anterhinum linaria.
" Weed.............	Benzoin odoriferum.
Snargel.................	*Aristolochia serpentaria.*
"	Arctostaphylos uva ursi.
Snowball...............	Viburnum opulus.
Soap Tree..............	Quillaya saponaria.
Soapwort...............	Saponaria officinalis.
Solomon's Seal..........	Convallaria polygonatum.
Sorrel, Sheep...........	Oxalis acetosella.
Sour Dock..............	Rumex crispus.
Southern Gentian.......	Gentiana catesbæi.
" Wood........	Artemisia abrotanum.

LOCAL OR COMMON NAMES.	BOTANICAL OR OFFICINAL NAMES.
Spanish Root	Glycyrrhiza glabra.
" Saffron	Crocus sativus.
Spatter Dock	Nuphar advena.
Spearmint	Mentha viridis.
Speedwell	Veronica officinalis.
Spice Berry	*Gaultheria procumbens.*
" "	Laurus benzoin.
" Birch	Betula lenta.
" Bush	*Benzoin odoriferum.*
" "	Laurus benzoin.
" Wood	Laurus benzoin.
Spicy Wintergreen	Gaultheria procumbens.
Spignet	*Aralia racemosa.*
"	Nard.
Spiked Loose Strife	Lythrum salicaria.
Spikenard	*Aralia racemosa.*
"	Nard.
Spindle Bush	Euonymus atropurpureus.

LOCAL OR COMMON NAMES.	BOTANICAL OR OFFICINAL NAMES
Spindle Tree...................	Euonymus atropurpureus.
Spleenwort...................	Asplenium angustifolium.
" Bush...............	Comptonia asplenifolia.
" Leaved Gale	" "
Split Rock........................	Heuchera Americana.
Spoonwood......................	Kalmia latifolia.
Spotted Alder....................	Hamamelis Virginica.
' Cardus................	Centaurea benedicta.
" Comfrey...............	Pulmonaria officinalis.
' " Cowbane..............	Cicuta maculata.
" Cranebill..............	Geranium maculatum.
" Geranium.............	" "
" Lungwort............	Pulmonaria officinalis.
" Thistle..................	Centaurea benedicta.
Square Stalk............	*Monarda didyma.*
" "	Scrophularia nodosa.
Squaw Mint..............	Hedeoma pulegioides.

LOCAL OR COMMON NAMES.	BOTANICAL OR OFFICINAL NAMES.
Squaw Root....................	*Orobanche uniflora.*
" "	Caulophyllum thalictroides.
" "	Cimicifuga racemosa.
" "	Macrotys racemosa.
" **Stalk**....................	Monarda didyma.
" **Vine**....................	Mitchella repens.
" **Weed**....................	*Senecio obovatus.*
" "	" aureus.
" "	Erigeron purpureum.
Staff Tree....................	Celastrus scandens.
" **Vine**....................	" "
Stagger Weed....................	Corydalis formosa.
Star Bloom....................	Spigelia marilandica.
" **Grass**....................	Aletris farinosa.
" **Root**....................	" "
" **Thistle**....................	Centaurea benedicta.
" **Wort**....................	Helonias dioica.

LOCAL OR COMMON NAMES.	BOTANICAL OR OFFICINAL NAMES.
Stavesacre	Delphinium staphisagria.
Steeple Bush	Spiræa tomentosa.
Stickwort	Agrimonia eupatoria.
Saint Johnswort	Hypericum perforatum.
Stillingia	Stillingia sylvatica.
Stinging Nettle	Urtica dioica.
Stinking Balm	Hedeoma pulegioides.
" Chamomile	Anthemis cotula.
" Pothos	Dracontium fœtidum.
Stink Weed	Datura stramonium.
Stinking Weed	Chenopodium anthelminticum
Stone Mint	Cunila mariana.
" Root	Collinsonia canadensis.
Storksbill	Geranium maculatum.
Stramonia	Datura stramonium.
Strawberry Leaves	Fragaria Virginiana.
" Tree	Euonymus atropurpureus.
Striped Alder	*Prinos verticillatus.*

LOCAL OR COMMON NAMES.	BOTANICAL OR OFFICINAL NAMES.
Striped Alder	Hamamelis Virginica.
Succory, Wild.	Centaurea Americana.
Sumach, Pennsylvania	Rhus glabrum.
" Smooth	" "
" Upland	" "
Summer Savory	Satureia hortensis.
Suter Berry	*Zanthoxylum fraxineum.*
" "	Aralia nudicaulis.
Swallow Root	Asclepias tuberosa.
Swallowwort	" "
Swamp Alder	Alnus rubra.
" Cabbage	Dracontium fœtidum.
" Dogwood	Cornus sericea.
" Hellebore	Helleborus Americana.
" Laurel	*Kalmia glauca.*
" "	Magnolia glauca.
" Magnolia	" "
" Milkweed	Asclepias incarnata.
" Sassafras	Magnolia glauca.
" Spleenwort	Asplenium angustifolium.

LOCAL OR COMMON NAMES.	BOTANICAL OR OFFICINAL NAMES.
Sweat Root	Polemonium reptans.
Sweating Plant	Eupatorium perfoliatum.
Sweet Balm	*Melissa officinalis.*
" "	Dracocephalum canariense.
" Balsam	Gnaphalium polycephalum.
" Basil	Ocymum basilicum.
Sweet Bay	*Magnolia glauca.*
" "	Laurus nobilis.
" Birch	Betula lenta.
" Bitter	Triosteum perfoliatum.
" Bugle	Lycopus Virginicus.
" Bush	Comptonia asplenifolia.
" " Ferry	" "
" Cicily	Osmorrhiza brevistylis.
" Clover	Melilotus officinalis.
" Elder	Sambucus canadensis.
" Elm	Ulmus fulva.
" Fern	Comptonia asplenifolia.
" " Bush	" "

LOCAL OR COMMON NAMES.	BOTANICAL OR OFFICINAL NAMES.
Sweet Flag	Acorus calamus.
" Flag Root	Calamus aromaticus.
" Gale	Myrica gale.
" Magnolia	Magnolia glauca.
" Marjoram	Origanum majorana.
" Rush	Acorus calamus.
" Scabious	Erigeron Philadelphicum.
" Scented Golden Rod	Solidago odora.
Sweet-scented Water Lily	Nymphæa odorata.
Sweet Willow	Myrica gale.
" Wood	Glycyrrhiza glabra.
Tackamahak	Populus balsamifera.
Tag Alder, Red	Alnus rubra.
Tall Cone Flower	Rudbeckia laciniata.
" Veronica	Leptandria Virginica.
Tansy	Tanacetum vulgare.
" Double	" crispus.

LOCAL OR COMMON NAMES.	BOTANICAL OR OFFICINAL NAMES.
Tea, Mountain	Gaultheria procumbens.
Tea Berry	" "
Tetterwort	*Chelidonium majus.*
" "	Sanguinaria canadensis.
Texas Sarsaparilla	Menispermum "
Thimble Berry	Rubus occidentalis.
" Weed	Rudbeckia laciniata.
Thistle, Blessed	Centaurea benedicta.
" Root	" "
Thorn Apple	Datura stramonium.
Thorough Root	Eupatorium perfoliatum.
" Stem	" "
" Wax	" "
" " Stem	" "
" Wort	" "
" " Root	" "
Thousand Leaves	Achillea millefolium.

LOCAL OR COMMON NAMES.	BOTANICAL OR OFFICINAL NAMES.
Three-leaved Spirea.........	Gillenia trifoliata.
Throat Root	*Geum rivale.*
" "	" Virginianum.
Thyme...............................	Thymus vulgaris.
Thymus............................	" "
Tick Weed........................	Hedeoma pulegioides.
Toad Flax.........................	Antirrhinum linaria.
" Lily	Nymphæa odorata.
Tobacco, Indian...............	Lobelia inflata.
Tochillies.........................	Capsicum annuum.
Toothache Bush...............	*Zanthoxylum fraxineum.*
" "	Aralia spinosa.
" Tree....................	" "
Tormentilla.......................	Geranium maculatum.
Touch Me Not..................	Impatiens balsamina.
Trailing Arbutus..............	Epigæa repens.
" Gaultheria..........	Gaultheria procumbens.

LOCAL OR COMMON NAMES	BOTANICAL OR OFFICINAL NAMES.
Traveler's Joy..........	Clematis Virginica.
Tree Primrose..........	Œnothera biennis.
Trembling Poplar.......	Populus tremuloides.
True Love..............	Trillium pendulum.
" Unicorn Root......	*Chamælirium luteum.*
" " "	Aletris farinosa.
Trumpet Weed.........	*Eupatorium purpureum.*
" "	Bignonia leucoxylon.
Tuber Root............	Asclepias tuberosa.
Tulip Bearing Poplar......	Liriodendron tulipifera.
" Tree Bark............	" "
Turkey Corn............	*Corydalis formosa.*
" "	Tephrosia corydalis.
" Pea............	*Corydalis formosa.*
" "	Tephrosia Virginiana.
Turmeric............	*Curcuma longa.*
"	Hydrastis canadensis.
Turnip, Indian............	Arum triphyllum.
" Wild............	" "
Turtle Bloom............	Chelone glabra.

LOCAL OR COMMON NAMES.	BOTANICAL OR OFFICINAL NAMES.
Turtle Head....................	Chelone glabra.
Umbil Root....................	Cypripedium pubescens.
Uncum	Senecio aureus.
Unicorn's Horn..............	Helonias dioica.
Unsteetle	Spigelia marilandica.
Upland Cranberry.......	Arctostaphylos uva ursi.
" Sumach........	Rhus glabra.
Uva Ursi...............	Arctostaphylos uva ursi.
Valerian....	Valeriana officinalis.
" American......	Cypripedium pubescens.
" Greek.........	Polemonium reptans.
Various-leaved Fleabane	Erigeron heterophyllum.
Vegetable Antimony.......	Eupatorium perfoliatum.
" Sulphur...........	Lycopodium clavatum.
Velvet Leaf...................	Cissampelos pareira.
Veratrum Viride..............	Helleborus Americana.
Verbena...........	Verbena officinalis.
" Nettle-leaved......	" urticifolia.
Vermont Snake Root........	Asarum canadense.

LOCAL OR COMMON NAMES.	BOTANICAL OR OFFICINAL NAMES.
Veronica	Leptandra Virginica.
Vervain	Verbena officinalis.
" **Nettle-leaved**	" urticifolia.
" **Blue**	" hastata.
Vine Maple	Menispermum canadense.
Violet Bloom	Solanum dulcamara.
Virginia Catch Fly	Silene Virginica.
" **Hoarhound**	Lycopus Virginicus.
" **Snake Root**	Aristolochia serpentaria.
" **Speedwell**	*Veronica officinalis.*
" "	Leptandra Virginica.
" **Winter Berry.**	Prinos verticillatus.
Virgin's Bower	Clematis Virginica.
Wafer Ash	Ptelea trifoliata.
Waahoo	Euonymus atropurpureus.
Wake Robin	*Arum maculatum.*
" "	Trillium latifolium.

LOCAL OR COMMON NAMES.	BOTANICAL OR OFFICINAL NAMES.
Wandering Milk Weed....	Apocynum androsæmifolium.
Water Avens.......................	Geum rivale.
" **Bugle**	Lycopus Europeus.
" **Cabbage**....................	Nymphæa odorata.
" **Carrot**.......................	Daucus carota.
" **Clover**.......................	Trifolium repens.
" **Cress**.......................	Nasturtium amphibium.
" **Cup**.......................	Saracenia purpurea.
" **Dock**	Rumex aquaticus.
" **Eryngo**	Eryngium aquaticum.
" **Flag**	Iris versicolor.
" **Hemlock**	*Conium maculatum.*
" "	Cicuta virosa.
" "	Œnanthe phellandrium.
" **Hoarhound**.............	Lycopus Virginicus.
" "	" Europeus.
" **Lily**............................	Nymphæa odorata.
" **Melon Seed**.............	Cucurbita citrullus.

LOCAL OR COMMON NAMES.	BOTANICAL OR OFFICINAL NAMES
Water Nymph.....................	Nymphæa odorata.
" Parsnip.................	*Sium nodiflorum.*
" "	Cicuta virosa.
" Pepper.................	*Polygonum punctatum.*
" "	" hydropiper.
" Plantain	Alisma plantago.
" Radish.....................	Nasturtium amphibium.
" Shamrock................	Menyanthes trifoliata.
Wax Berry.......................	Myrica cerifera.
" Myrtle......................	" "
" Work.........................	Celastrus scandens.
Way Bread.........................	Plantago major.
White Ash Bark........	Fraxinus acuminata.
" Avens............	Geum Virginianum.
" Bay...............	Magnolia glauca.
" Balsam...........	Gnaphalium polycephalum.
" Bane Berry......	Actæa alba.
" Beads............	" "
" Canella...........	Canella "

Local or Common Names.	Botanical or Officinal Names.
White Cap	Spiræa tomentosa.
" Clover	Trifolium pratense.
" Cohosh	Actea alba.
" Flux Root	Asclepias tuberosa.
" Gentian	Triosteum perfoliatum.
" Hellebore	Veratrum album.
" Hoarhound	Marrubium vulgare.
" Indian Hemp	Asclepias incarnata.
" Leaf	*Spirea tomentosa.*
" "	Chimaphilla maculata.
" Lettuce	Nabulus albus.
" Oak Bark	Quercus alba.
" Pond Lily	Nymphæa odorata.
" Poplar	*Liriodendron tulipifera.*
" "	Populus tremuloides.
" Root	Asclepias tuberosa.
" Sanicle	*Aristolochia serpentaria.*

Local or Common Names.	Botanical or Officinal Names.
White Sanicle	Eupatorium aromaticum.
" Snake Root	*Eupatorium urticifolium.*
" "	Aristolochia serpentaria.
" Tubrous-rooted Swallowwort	Asclepias tuberosa.
" Vervain	Verbena urticifolia.
" Walnut	Juglans cinerea.
" Willow	Salix alba.
" Wood	Liriodendron tulipifera.
Wickup	Epilobium angustifolium.
Wild Allspice	Laurus benzoin.
" Angelica	*Angelica sylvestris.*
" "	Angelica archangelica.
" Archangel	" "
" Basil	Cunila mariana.
" Celandine	Impatiens fulva and pallida.
" Carrot	Daucus carota.
" Chamomile	Anthemis cotula.

LOCAL OR COMMON NAMES.	BOTANICAL OR OFFICINAL NAMES.
Wild Cinnamon	Canella alba.
" Cherry Bark	Prunus Virginiana.
" Coffee	Triosteum perfoliatum.
" Columbo	Frasera walteri.
" Cranberry	Arctostaphylos uva ursi.
" Dwarf Elder	Aralia hispida.
" Ginger	Asarum canadense.
" Hydrangea	Hydrangea arborescens.
" Hyssop	Hyssopus officinalis.
" Indigo	Baptisia tinctoria.
" Ipecac	*Triosteum perfoliatum.*
" "	Euphorbia ipecacuanha.
" Jalap	Convolvulus panduratus.
" Jessamine	Gelseminum sempervirens.
" Lemon	Podophyllum peltatum.
" Lettuce	*Pyrola rotundifolia.*
" "	Lactuca sativa.

LOCAL OR COMMON NAMES.	BOTANICAL OR OFFICINAL NAMES.
Wild Liquorice	Smilax officinalis.
" Mandrake	Podophyllum peltatum.
" Marjoram	Origanum majorana.
" Pink	Silene Virginica.
" Potato	Convolvulus panduratus.
" Red Raspberry	Rubus strigosus.
" Rhubarb	Convolvulus panduratus.
" Sage	Salvia lyrata.
" Sarsaparilla	Aralia nudicaulis.
" Scammony	Convolvulus panduratus.
" Senna	Cassia marilandica.
" Snowball	Ceanothus Americana.
" Succory	*Sabbatia angularis.*
" "	Centaurea Americana.
" Tobacco	Lobelia inflata.
" Turnip	*Arum triphyllum.*
" "	Asarum canadense.
" Wormwood	Artemisia vulgare.
" Yam	Dioscorea villosa.
Willow Bark	Salix alba.

Local or Common Names.	Botanical or Official Names.
Willow Herb, Hooded......	Scutellaria lateriflora.
" Rose.................	Cornus sericea.
Wind Flower.....................	Anemone pratensis.
" Root.....................	Asclepias tuberosa.
Windwort........................	" "
" Root..................	" "
Wing Seed.....................	Ptelea trifoliata.
Winter Berry..................	Prinos verticillatus.
" Bloom......................	Hamamelis Virginica.
" Clover.....................	Mitchella repens.
" Fern.................	Pteris atropurpurea.
Wintergreen.....................	Gaultheria procumbens.
" American.....	Chimaphila umbellata.
Winter Pink.............	Epigæa repens.
Witch Grass.....................	Triticum repens.
Witch Hazel.................	Hamamelis Virginica.
Wolfbane......................	Veratrum viride.

LOCAL OR COMMON NAMES	BOTANICAL OR OFFICINAL NAMES.
Wolfsbane	*Aconitum napellus.*
Wood Betony	Pedicularis canadensis.
" Sorrel	Oxalis acetosella.
Woodbine	Gelsiminum sempervirens.
Woody Nightshade	Solanum dulcamara.
Worm Goose Foot	Chenopodium anthelminticum.
" Grass	Spigelia marilandica.
" Seed	*Chenopodium anthelminticum.*
" "	Spigelia marilandica.
" Wood	Artemisia absinthium.
Wound Wort	Collinsonia canadensis.
Yarrow	Achillea millefolium.
Yaw Root	Stillingia sylvatica.
Yellow Adder's Tongue	Erythronium Americanum.
" Broom	Baptisia tinctoria.
" Bugle	Ajuga chamæpithys.
" Dock	Rumex crispus.
Yellow-flowered Rhodo-dendron	Rhododendron chrysanthemum.

LOCAL OR COMMON NAMES.	BOTANICAL OR OFFICINAL NAMES.
Yellow Gentian	*Frasera walteri.*
" "	Gentiana lutea.
" **Jessamine**	Gelseminum sempervirens.
" **Melilot**	Melilotus officinalis.
" **Moth**	Verbascum thapsus.
" **Parilla**	Menispermum canadense.
" **Pond Lily**	Nuphar advena.
" **Poplar**	Liriodendron tulipifera.
" **Puccoon**	Hydrastis canadensis.
" **Root**	" "
" "	Coptis trifolia.
" "	Zanthoriza apiifolia.
" **Snake Leaf**	Erythronium Americanum.
" **Toad Flax**	Anterrhinum linaria.
" **Weed**	Ranunculus acris.
" **Wood**	*Zanthoxylum fraxineum.*
" "	Aralia nudicaulis.
Yerba Santa	Artemisia vulgare.
Zanthoriza	Rumex aquaticus.
Zedoary Root	Zedoary radix.

ALPHABETICAL LIST

OF THE

COLLEGES OF PHARMACY

IN THE

UNITED STATES,

WHOSE DIPLOMAS ARE GRANTED ONLY WHEN THE STUDENT POSSESSES, IN ADDITION TO THE THEORETICAL OR SCIENTIFIC KNOWLEDGE ACQUIRED BY STUDY, A PRACTICAL ACQUAINTANCE WITH THE APOTHECARY BUSINESS, OBTAINED BY ACTUAL EXPERIENCE FOR SEVERAL YEARS PREVIOUS TO EXAMINATION.

CHICAGO COLLEGE OF PHARMACY,	Chicago, Ill.
CINCINNATI COLLEGE OF PHARMACY,	Cincinnati, O.
LOUISVILLE COLLEGE OF PHARMACY,	Louisville, Ky.
MARYLAND COLLEGE OF PHARMACY,	Baltimore, Md.
MASSACHUSETTS COLLEGE OF PHARMACY,	Boston, Mass.
NEW YORK COLLEGE OF PHARMACY,	New York, N. Y.
PHILADELPHIA COLLEGE OF PHARMACY,	Philadelphia, Pa.
ST. LOUIS COLLEGE OF PHARMACY,	St. Louis, Mo.

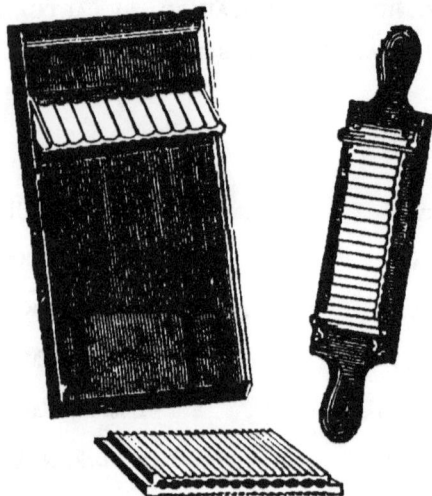

P. W. ENGS & SONS

WINE MERCHANTS,

131 Front Street, New-York.

CORNER OF PINE.

————————◆————————

BRANDIES, IMPORT

From the old houses of OTARD, DUPUY & Co., J. HENNESSY & Co., and J. & F. MARTELL, and PINET, CASTILLON & Co., Cognac; PELLEVOISIN FRERES, La Rochelle, France; also, selections by their friends in London, from the Docks; in Half Pipes, Quarter, Fifth and Eight *Casks*.

Also: The largest Assortment of grades from J. HENNESSY & Co., J. & F. MARTELL, OTARD, DUPUY & Co., and PINET, CASTILLON & Co., to be found in the U. S. in *Cases*.

JAMAICA RUMS,

The best selections from London Docks, of high flavor and proof, in puncheons, hhds., bbls. and *Half Bbls*.

ST. CROIX RUMS,

In puncheons and barrels, celebrated " L B " brands, both new and very old.

" MEDER SWAN GIN,"

In half pipes, bbls. and *Half Bbls*., at Agents' prices.

SCOTCH MALT WHISKEY,

J. RAMSEY & ARDBEG, in pun's, hf. pun's, bbls. and *Half Bbls*.

IRISH MALT WHISKEY, CORRIGAN & Co.

From the celebrated " Ballymony " Distillery. in half pun's and bbls.

AGENTS in the UNITED STATES for OLD TOM GIN,

From Sir R. BURNETT & Co., London, in pun's, half pun's, bbls. and *Cases*.